The Path to a Sustainable Civilisation

"It's situation critical and time's running out. Here's a book that, without indulging in blind optimism, crisply sets out solutions that can work if we get on with it immediately."
—Kerryn Higgs, University Associate, University of Tasmania, Associate Fellow of the Club of Rome, and author of *Collision Course: Endless Growth on a Finite Planet*

"*The Path to a Sustainable Civilisation* correctly identifies the dominant economic system as one of the fundamental drivers of the exploitation of our planet's environment and the vast majority of its people. The book recommends ending economic growth by the rich countries and transitioning to a steady-state economy with sustainable prosperity for all, guided by the transdisciplinary framework of ecological economics and Modern Monetary Theory. In presenting its case, this book is readable, well documented and convincing."
—Steve Keen, Honorary Professor and Distinguished Research Fellow, Institute for Strategy, Resilience and Security, University College London

Mark Diesendorf • Rod Taylor

The Path to a Sustainable Civilisation

Technological, Socioeconomic
and Political Change

palgrave
macmillan

Mark Diesendorf (iD)
School of Humanities & Languages
UNSW Sydney
Sydney, NSW, Australia

Rod Taylor
Canberra, ACT, Australia

ISBN 978-981-99-0662-8   ISBN 978-981-99-0663-5   (eBook)
https://doi.org/10.1007/978-981-99-0663-5

This Palgrave Macmillan imprint is published by the registered company Springer Nature Singapore
Pte Ltd.
The registered company address is: 152 Beach Road, #21-01/04 Gateway East, Singapore 189721,
Singapore

Dedicated to earth system scientist, Will Steffen, 1947–2023

Acknowledgements

We thank Margot Nash, for reading several chapters and suggesting valuable improvements in expression and readability, and for encouragement. We thank Anne Taylor for continuing support and guidance. We express our appreciation to Marion Duval, Commissioning Editor of Palgrave Macmillan, who originally perceived the value of our book concept and was strongly supportive throughout the writing process. We received valuable comments on various topics discussed in the book from Steven Bunce, John Diesendorf, Joseph Diesendorf-Young, Alison Field, Steven Hail, Manfred Lenzen, Brian Martin, John Quiggin and Stephen Williams. We thank Connie Li of Springer Nature for her diligent editorial work.

The Stockholm Resilience Centre at Stockholm University kindly granted us permission to reproduce Fig. 2.1 on planetary boundaries.

Contents

About the Authors

Mark Diesendorf is Honorary Associate Professor in the Environment & Society Group, School of Humanities & Languages at UNSW Sydney. Originally trained as a physicist, he broadened out into interdisciplinary energy and sustainability research. From 1996 to 2001 he was Professor of Environmental Science and Founding Director of the Institute for Sustainable Futures at University of Technology Sydney. His previous books include *Sustainable Energy Solutions for Climate Change* (2014), *Climate Action: A campaign manual for greenhouse solutions* (2009), *Greenhouse Solutions with Sustainable Energy* (2007); and *Human Ecology, Human Economy: Ideas for an ecologically sustainable future* (co-editor, 1997).

Rod Taylor is a freelance science and technology writer, journalist and broadcaster. His book, *Ten Journeys on a Fragile Planet* (2020), has received strong positive reviews and been the subject of numerous public appearances. He is co-editor of the new book, *Sustainability and the New Economics* (Springer, 2022). His weekly science column and other articles have been published in Fairfax and now Australian Community Media masthead papers for over 13 years. One of his columns was featured in *Best Australian Science Writing* (2018). He has also written several pieces broadcast on national radio by the Australian Broadcasting Corporation.

The authors' website for the book is https://sustainablecivilisation.com.

Abbreviations

100RE	100% Renewable Energy
100RElec	100% Renewable Electricity
AEMO	Australian Energy Market Operator
ALP	Australian Labor Party
ASIO	Australian Security Intelligence Organisation
CO_2	Carbon Dioxide
CST or CSP	Concentrated Solar Thermal Power
EE	Energy Efficiency
EROI or EROEI	Energy Return on Energy Invested
ESD	Ecologically Sustainable Development
ETS	Emissions Trading Scheme
EV	Electric Vehicle
FF	Fossil Fuel
FSC	Forest Stewardship Council
GDP	Gross Domestic Product
GFC	Global Financial Crisis
GHG	Greenhouse Gas
GPI	Genuine Progress Indicator
GW	Gigawatt, Equals 1000 Megawatts (MW)
ICE	Internal Combustion Engine
IEA	International Energy Agency
IPCC	Intergovernmental Panel on Climate Change
ISEW	Index of Sustainable Economic Welfare

JG	Job Guarantee
LNG	Liquefied Natural Gas
MMT	Modern Monetary Theory
NATO	North Atlantic Treaty Organisation
NGO	Non-government Organisation
NPT	(Nuclear) Non-Proliferation treaty
NREL	National Renewable Energy Laboratory (USA)
OECD	Organisation for Economic Cooperation and Development
P2X	Power-to-X
PV	Photovoltaic
QE	Quantitative Easing
RE	Renewable Energy
RElec	Renewable Electricity
REZ	Renewable Energy Zone
SD	Sustainable Development
SDG	Sustainable Development Goal
SSE	Steady-State Economics or Economy
TFEC	Total Final Energy Consumption
TWh	Terawatt-Hour, Equals 1000 GWh
UBI	Universal Basic Income
UBS	Universal Basic Services
UK	United Kingdom
UN	United Nations
UNFCCC	United Nations Framework Convention on Climate Change
USA	United States of America
VRElec	Variable Renewable Electricity

List of Figures

List of Tables

List of Boxes

1

Introduction

A civilization that proves incapable of solving the problems it creates is a decadent civilization.

A civilization that chooses to close its eyes to its most crucial problems is a sick civilization.

A civilization that plays fast and loose with its principles is a dying civilization.

(Aimé Césaire)[1]

In April 1912, the ocean liner Titanic was on its maiden voyage across the Atlantic. Constructed with state-of-the-art technology, it was the largest ship ever built and thought to be unsinkable. The wealthy passengers were enjoying the luxurious first-class facilities; lower deck passengers fared less well; we can only imagine the almost hellish experiences of the workers stoking the massive boilers. All passengers and crew were oblivious to their impending doom. Most were unaware that, under pressure from wealthy industrialist owners, the captain was driving the ship at high speed through freezing waters, despite warnings of sea-ice. This was common practice at the time.

On 15 April, shortly before midnight, the ship struck an iceberg and, 2 hours and 40 minutes later, sank to the ocean floor nearly four

© The Author(s), under exclusive license to Springer Nature Singapore Pte Ltd. 2023
M. Diesendorf, R. Taylor, *The Path to a Sustainable Civilisation*,
https://doi.org/10.1007/978-981-99-0663-5_1

kilometres below the surface. About 1500 passengers and crew perished, including three-quarters of third-class passengers, just over half of second-class and 38% of first-class.

Today Spaceship Earth and its passengers share many similarities with the Titanic such as the complexity of a trusted system capable of spectacular failure; the warnings ignored; the dominance of economic drivers; the power imbalance among those on board and the increased risks faced by the powerless; the lack of respect for the natural environment; and the attitudes, values and vested interests of the people in charge.

While there is obviously no risk of Earth colliding with an iceberg, collision with a stray asteroid could happen in the distant future. But, as we write these words, civilisation faces immediate existential threats— threats caused by us. Human activities are changing the global climate: melting ice sheets, glaciers and permafrost; lifting sea-levels; changing the major ocean currents; and extinguishing species. Millions of people around the world are suffering the impacts of human-induced climate change manifest in the increased frequency of droughts, heatwaves, wildfires and floods. If current trends continue, large regions of the Earth could become uninhabitable by people. In addition to driving climate change, human activities are converting forests to pasture; turning fertile soils into deserts and suburbs; polluting the atmosphere, land and freshwater; and decimating biodiversity.

Meanwhile, the very resources needed to combat the changes are being diminished. Among these are the mineral resources needed to transition energy systems away from fossil fuels; the verdant forests, mangroves and phytoplankton that are the lungs of the biosphere; the fertile soils necessary for food production; and the unpolluted waterways essential for continuing life.

These trends are caused by increasing consumption and population, polluting technologies, and an economic system based on exploiting the natural environment and people, a system that refuses to recognise any limits to growth on a finite planet. Driving these trends are powerful multinational corporations, other organisations with vested interests, and politicians whose goals are short-term and self-serving.

In addition to the environmental threats is the threat of nuclear annihilation, increasing in the short-term while the cruel war rages in Ukraine; increasing in the longer term as countries use nuclear power program as a cloak to hide their development of nuclear weapons.

As our spaceship forges onwards, the divisions between its first-, second- and third-class passengers are widening. The rich are getting richer, yet little if anything is trickling down to the poor. Globalisation in its current form gives great power to multinational corporations to strip the poor nations of their forest, fish and mineral resources, giving low returns to local custodians, undermining the potential for these countries to develop their own economies and leaving the local people with precarious incomes. Low-income and indigenous people have the worst health, even in the rich countries. Women still struggle for wage justice and access to power, even in the richest countries. In some parts of the world, women are still treated like chattels of the men; forbidden education beyond primary school, under-paid in permitted jobs, or excluded entirely from jobs and forced to remain hidden from public view. Slavery, although banished officially by governments, continues unofficially in various forms. Racism and sexism remain covert drivers of global patriarchal systems of power.

Disillusionment with the system of representative government, resulting from its failure in many countries to address the needs of the poor and disadvantaged, is driving a drift to blind faith in autocrats and powerful religious leaders. These rulers enhance their power by hobbling political opponents and independent media, compromising the judiciary, reversing progressive legislation, and creating a secret police force loyal to them and/or building a *de facto* dictatorship. Authoritarian states—like Russia, China, Egypt and Turkey—generally keep a façade of democracy while restricting opposition. Government decision-making in nominally 'democratic' states, such as the United States of America, is controlled to a large degree by powerful interest groups—especially business, industry and the military[2]; they impose their economic, political and cultural ideologies on many other countries.

These are the challenges we must face. Since time is of the essence, we must act very rapidly, focusing on the key issues. Then it may still be possible to mitigate the existential threats and transition to a better society that is ecologically sustainable, more socially just and less warlike. We call this goal 'the Sustainable Civilisation'. Our book discusses the strategies and policies needed for transitioning from the present collapsing civilisation to the sustainable one.

Unlike some future social scenarios, we do not assume that the new society is built from the shattered fragments of the old. This is a possible interpretation of the statement by novelist Sally Rooney: "*Or maybe it was just the end of one civilization, ours, and at some time in the future another will take its place.*"[3] We believe instead that technological change and socioeconomic change can and must be implemented now to save the planet and its inhabitants. We also reject the scenario in which the present global population of eight billion transitions to millions of small, self-sufficient communities living off the land. Even with half the present global population, such a society would rapidly destroy the environment upon which it depends.

The scenario proposed in this book is one of hope, hope for a transition with treacherous gaps in the pathway that can still be repaired. It starts from the present civilisation with all its faults, dangers and threats. It embraces the very real possibility of implementing widespread technological change together with necessary, fundamental changes to the economic system and the system of governance.

The Sustainable Civilisation is not a utopia but, unlike the present collapsing civilisation, it offers the possibility of the human experiment lasting well beyond the present century and into the next millennium. It has the best of modern, clean technologies envisaged in the Green New Deal,[4] while recognising that global growth in consumption and population cannot continue without environmental, social and economic collapse. It has a new economic system that is designed for environmental protection and restoration, and social justice. It has strengthened democratic systems of government. The transition demands a big program, but a necessary one, and offers hope that there is a future for humanity.

Chapter 2 documents the principal threats to the present civilisation sketched in this introduction and justifies concerns that the existing system is heading for collapse. It explains why we must solve both the environmental crises and declining social justice, pointing out that the rich countries and rich individuals have the principal responsibility for both. The subsequent chapters draw a clearer picture of the Sustainable Civilisation, put forward the key strategies for transitioning energy systems and material resources, examine the problem of state capture by corporate interests and how to overcome it, and discuss how we can transition from the present dominant economic system that fosters exploitation of the planet and its people to one that fosters ecological sustainability and social justice.

Chapter 3 explains the concepts of sustainability and sustainable development and then uses strong versions of them to characterise the Sustainable Civilisation. The rich countries, the Global North, have the resources to protect and restore the environment and improve social justice. Their necessary strategy is fairly clear, although the barrier of vested interests clinging tenaciously to business-as-usual is substantial. The Global South faces an even more formidable barrier, the struggle against the legacy of colonialism and the constraints of neo-colonialism. It must find future pathways that do not necessarily conform to the Western model of 'development'.

Because energy is one of the foundations of the modern industrial civilisation, Chap. 4 explains how an ecologically sustainable energy system, based on renewable energy and energy efficiency, can form the basis of the Sustainable Civilisation. We refute the principal myths disseminated by vested interests to undermine confidence in this technically feasible, affordable, energy future. We also show that, unless global energy consumption can be reduced, then it's unlikely that the transition can be completed in time to keep global heating well below 2°C, the Paris Agreement's climate change target.

Chapter 5 discusses the transitions needed for physical resources—minerals and other raw materials, forests, soils, food, freshwater and agriculture—and technologies. While renewable resources can be conserved by reducing their rates of use to their rates of regeneration, managing non-renewable resources is more difficult. Reduced consumption, substitution, reuse, recycling remanufacture and redesign can greatly extend their lifetimes, possibly for millennia, but ultimately some key non-renewable resources will become inaccessible due to economic and energy costs—a completely circular economy is impossible.

While Chaps. 4 and 5 develop strategies and policies specifically for energy and other natural resources, Chap. 6 discusses the roots of environmental destruction and social injustice at a more fundamental level, focusing on the capture of nation-states by corporate and other vested interests. For the Global South, state capture overlaps with neo-colonialism. The chapter discusses strategies for cutting these roots while strengthening democracy in both North and South. International action is needed as well as national.

Chapter 7 critiques one of major roots of the crisis, conventional economics, and proposes strategies for transitioning to the interdisciplinary field of ecological economics. It proposes that the rich countries could finance the transition to the Sustainable Civilisation by drawing upon the insights provided by Modern Monetary Theory.

Chapter 8 sets out a strategy for social change, actually two related strategies: citizens exert pressure on power-holders, while simultaneously building their own alternatives, to achieve the necessary socioeconomic and technological transformations. The chapter provides a compendium of government policies and actions by community groups needed to drive the transition. This chapter also refutes some of the myths, objections and obfuscations used to discourage social change movements—myths that are likely to be used against the proposed transition to the Sustainable Civilisation.

Chapter 9 concludes the book.

Notes

1. Aimé Césaire (1972, 2000). *Discourse on Colonialism*. Translated by Joan Pinkham. New York: Monthly Review Press, p.31. https://files.libcom.org/files/zz_aime_cesaire_robin_d.g._kelley_discourse_on_colbook4me.org_.pdf (Original version published by Éditions Réclame in 1950).

2. In his farewell speech on 17 January 1961, President Dwight D. Eisenhower issued his famous prescient warning to the American people about the growing power of the 'military-industrial complex'. https://www.politico.com/story/2019/01/17/eisenhower-warns-of-military-industrial-complex-jan-17-1961-1099265
3. Sally Rooney (2021). *Beautiful World, Where Are You*. Faber and Faber.
4. Green Party US website. https://www.gp.org/green_new_deal; https://www.greennewdealuk.org/

Reference

Weblinks accessed 26/10/2022.

2

The Sustainability Crisis

Human actions continue to push the planet towards its existential and ecosystem limits. (UN Global Assessment Report on Disaster Risk Reduction)[1]

2.1 Our Planet in Distress

Although conventional economists believe that we humans have risen above our animal origins, ecology and environmental science inform us that we are completely dependent on the natural environment. We depend on the atmosphere to give us a benign climate, oxygen for respiration, and protection from harmful ultraviolet radiation from the Sun. We depend on the global environment to provide the food, water and energy that we ingest and digest, the raw materials for our goods and services and, within limits, a waste management system. The Earth, or more precisely, its biosphere, is our life support system.

Yet the evidence is as overwhelming as the message is dire: the Earth is in crisis. For decades the growing warning signs have been ignored, attacked and deliberately distorted to the point that our best hope now is for a soft landing rather than a crash. While climate change is forefront

© The Author(s), under exclusive license to Springer Nature Singapore Pte Ltd. 2023
M. Diesendorf, R. Taylor, *The Path to a Sustainable Civilisation*,
https://doi.org/10.1007/978-981-99-0663-5_2

in the popular mind, it is only one of many threats that are approaching from multiple, interconnected directions: they include loss of biodiversity, air pollution, stratospheric ozone depletion, deforestation, overuse and pollution of freshwater, loss of soils and desertification.

Biodiversity helps maintain our soils and water quality, pollinates plants and crops and disperses their seeds, decomposes organic wastes, provides some of the ingredients of medicines, and supplies food, fibre and essential nutrients. Ambient and household air pollution is responsible for about seven million deaths each year, according to the World Health Organisation[2]—diseases include heart disease, stroke, chronic obstructive pulmonary disease, cancer and pneumonia. Depletion of the ozone layer results in more exposure to ultraviolet B (UVB) radiation that is responsible for skin cancer and cataracts in humans, damage to plant growth and marine ecosystems, and more. Forests help maintain the oxygen levels in the air that we breathe, regulate our climate, provide habitats for biodiversity, improve water quality, reduce run-off and resulting soil loss during heavy rains and floods, and provide us with timber.

Planetary Boundaries

A valuable perspective on the environmental crisis is given by the Planetary Boundaries framework illustrated in Fig. 2.1. The areas inside the dotted-line boundary indicate the safe operating limits, beyond which the system becomes unstable and therefore likely to crash. If the concept were applied to, say, a car, the boundaries would include things such as engine temperature, speed, oil and water levels, and mechanical condition. And, since the car is operated by a person, we ought to consider whether the driver is able to maintain control. While the Planetary Boundaries approach does not explicitly include humans, this is a critical component of our present civilisation and a future ecologically sustainable civilisation. As Fig. 2.1 shows, we have now transgressed the safe operating space in six of the nine main categories of environmental impacts and almost reached it with a seventh. Research is continuing on the status of other environmental impacts. 'Biogeochemical Flows' includes the oxygen, carbon dioxide and phosphorus cycles; 'Novel Entities' still has to be disaggregated.

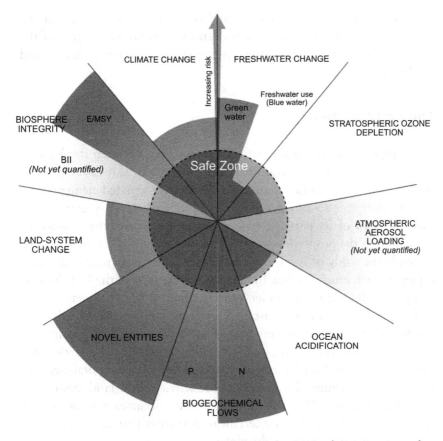

Fig. 2.1 Planetary boundaries as understood in 2022. (Source: Azote for Stockholm Resilience Centre/Stockholm University [3] Illustration based on analysis in Wang-Erlandsson et al. (2022)[4] and Steffen et al. (2015).[5] Note: E/MSY is Extinctions/Mammal Species Years; the biogeochemical flows beyond the safe operating limits are nitrogen (N) and phosphorus (P). Some sectors are not yet quantified. Updated Planetary Boundaries Figure from Wang-Erlandsson et al. 2022)

Taken individually, each of these impacts poses an existential threat to human civilisation, but what makes a truly wicked brew is that they are not independent but interact with one another. For example, climate change causes land system change—including loss of biodiversity, forests and soils, and freshwater change—and ocean acidification. Conversely, climate change is impacted by some land system changes. Most of the impacts of climate change have adverse effects on human health and wellbeing.

The Earth System

The complexity of the whole and the interactions between its elements mean that we are trying to understand a system, in this case the Earth System. A system is a group of interacting or interrelated elements that act according to a set of rules to form a unified whole. A system is more than the sum of its parts. It has 'emergent' properties that cannot be derived from the properties of its individual elements taken in isolation.

While the study of Earth System Science is relatively recent, the concept goes back much further. In Western science, one of the first thinkers along these lines was Alexander von Humboldt,[6] whose frenetic travels across South America and Eastern Europe crystallised the idea that the Earth is a deeply interconnected set of parts, where each affects the others. Geology affects soil types and water flow, which affect vegetation and therefore farming. Von Humboldt observed how changing drainage patterns could fundamentally affect the productivity of the land, both natural and agricultural. He even recognised that volcanoes are not isolated features, but rather reflect subterranean structures that extend, not just across continents, but across the globe.

Many Indigenous cultures have long been aware that their presence on the land affects its functioning, and that their future is intimately connected to its health. They see themselves as living *in* the land, not *on* it, and the modern notion that a person can own the land is in conflict with cultural traditions of collective stewardship. Aboriginal Australians say, "We don't own the land, the land owns us".

The contrast between systems thinking and 'point solutions' becomes obvious in the case of renewable energy, which we discuss in detail in

Chap. 4. A simplistic approach focuses on solar panels and wind turbines, while ignoring the fact that these are part of a grid where there are many suppliers and many consumers. An electricity grid, in which most electricity is generated by variable renewables, needs some form of storage and the means to manage technical aspects such as voltage and frequency. That grid then operates within a market where there are buyers and sellers, and marketing and accounting systems. From a consumer's perspective, electricity almost magically appears at a power socket, hiding the enormously complicated systems needed to deliver it.

At a much higher level, civilisation itself is the product of a vastly complicated arrangement of systems that supply, not just electricity, but food, education, housing, social security and consumer goods, among others. In short, civilisation provides all the goods and services expected by a citizen.

Systems thinking is the basis of ecology, the science of how organisms interact with one another and with their environment. While ecologists have understood the importance of systems for decades, others generally focus on individual elements of a system, for example, extinction of a species. However unfortunate, the full impact of the loss of a species can be understood only by considering the significance of the species in the ecosystem. While reductionist thinking emphasises the parts, systems thinking cares about connections, about how the many parts fit and work together. The real impact of extinction and habitat destruction is that every loss is another unwinding of the Earth's, our, life support system. We are removing bolts from its machinery at an ever-increasing rate, and at some point, it will fail. In fact, as we describe in this chapter, parts of the system are already failing.

Of particular concern is the relative timescales of humans and the Earth System. Humans typically think and act in terms of days, months or years, and only on rare occasions in terms of decades or longer. The climate system, however, responds over decades, centuries and more.

Climate Change

Arguably the failure needing the most urgent action is anthropogenic (human-induced) climate change. Earth's climate is an extremely complex system influenced by many elements, including solar energy input to the upper atmosphere, composition of the atmosphere, interaction between atmosphere and ocean, land use, reflectivity of Earth's surface, composition of soils, and the numbers, types and geographic distributions of living organisms. Human activity is interfering with all except the first of these elements.

Until recent years, climate change was seen as a future threat, but the impacts are already arriving more frequently and more severely than expected. Since 2000, the most devastating impacts have been heatwaves, droughts, wildfires and floods in many parts of the world. As the 2022 IPCC report[7] describes in its Fig. SPM.2b, every major facet of human existence is already being affected. In each case, the result has been 'negative' (i.e. destructive) or at best neutral. Recent research finds that global heating of 1.5°C could drive the Earth system beyond several tipping points into self-perpetuating, irreversible changes. The greater the heating, the greater the risk of crossing multiple tipping points.[8] On the current trend, global heating is 1.04°C above the pre-industrial average (1880–1900)[9] and could exceed 1.5°C within a decade en route to 2–3°C before the century's end.

I = PAT

People sometimes argue about whether the principal driver of environmental impact is the growth in affluence, in population or in polluting technology. An illuminating answer was given by environmental scientist Paul Ehrlich and physicist John Holdren[10]: *it is all three*. They showed that, at one conceptual level, environmental impact I can be written as population P multiplied by affluence A (consumption per person) multiplied by technological impact T, the well-known $I = PAT$ relationship. As shown in Box 2.1, this equation is identically true, that is, true because of the way A and T are defined.

Box 2.1 The Identity *I* = *PAT*

Let's measure affluence *A* by gross domestic product (GDP) per person and technological impact *T*, that is, how polluting a technology is, by environmental impact *I* per unit of GDP. Then *I* = *PAT* becomes:

$$I = P \times \left(\frac{GDP}{P} \right) \times \left(\frac{I}{GDP} \right),$$

where × denotes multiply. On the right-hand-side, the *P*s and GDPs cancel, and we are left with the identity $I = I$. It is true by definition.

Although the decomposition of environmental impact into the three factors is useful, as outlined in the main text, there are two problems in using it to describe the evolution of the system over time. The first is that *I* is in both the left-hand-side and the right-hand-side. The unknown future environmental impact is expressed in terms of itself. The other problem is the assumption that, if we double *P*, the factor GDP/*P* will remain constant. This is only true if GDP equals *P* multiplied by something that's independent of *P*. Similarly, if we double the affluence factor GDP/*P*, it is unrealistic to assume that the technology factor *I*/GDP remains constant.

There is debate as to whether the identity *I* = *PAT* is useful for describing the evolution of environmental impacts of a system over time. Some authors argue that it is preferable to have an equation with an unknown environmental impact on the left-hand-side and known or assumed driving forces on the right-hand-side. Then we can see more clearly the environmental impact of continuing growth in consumption, population, pollution and other factors.

Decomposing environmental impact *I* into the three factors *P*, *A* and *T* is useful because it shows explicitly that improving technologies alone may not be sufficient to protect the environment. For example, if we halve the technological impact *T* while the population doubles and affluence *A* remains the same, then environmental impact *I* is unchanged. Each factor requires different policies to reduce it and hence environmental impact. Which factor is more important? In practice, the rate of global population growth is gradually declining, while global consumption per person is, at the time of writing, returning to the pre-pandemic growth

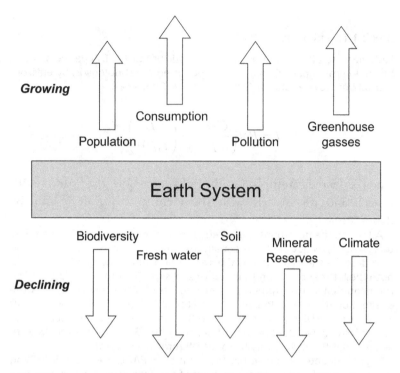

Fig. 2.2 Drivers of environmental impacts on the Earth system. (Note: Because of their importance, we have separated greenhouse gases from other forms of pollution in the figure)

rate, so the growth in affluence is of even greater concern than the growth in population.

It is really the combination of high affluence and population, that is, the population of the rich, that has the highest environmental impact. For example, a study carried out for Oxfam found that the carbon footprints of the world's richest 1% of people are 30 times greater than the level of per capita emissions consistent with the 1.5°C goal of the Paris Agreement's climate target.[11]

The environmental impact of the three drivers can be broken down into many specific impacts, as illustrated in Fig. 2.2.

Growth, Overshoot and Collapse

Scenario analysis can be valuable for developing policy interventions. The first, and possibly the most controversial, of such studies was published in 1972 by a group of researchers led by Donella Meadows at the Massachusetts Institute of Technology (MIT). The popular account is a book entitled *The Limits to Growth*.[12] The researchers used a computer model called World3 to examine the interactions between population, food production, industrial production, consumption of non-renewable resources and pollution. The model commenced with historical values of these variables from year 1900 to 1970 and from there explored several different possible future scenarios out to 2100.

It must be emphasised that scenarios are not predictions. A model is, by its very nature, a simplified version of reality. A scenario tells us, if we make certain assumptions, the likely outcomes. Confidence in scenario modelling is strengthened by using sensitivity analysis, that is, exploring how sensitive the results are to varying some of the assumptions. The MIT group did this.

The results showed that the 'business-as-usual' scenario, with continued growth in the global economy, led to 'overshoot', that is, exceeding the planetary limits, and that this would most likely lead to the collapse of the economic system and population around the mid-twenty-first century. The collapse resulted from diminishing resources and increasing environmental damage due to pollution. This result was found before climate change was recognised as a major threat.

In contrast, the 'stabilised world' scenario implemented both technological changes and social policies to stabilise population, material wealth, food and services per capita. This strategy suggested that collapse could be avoided.

An intermediate scenario attempted to achieve ecological sustainability by purely technological solutions. It succeeded in delaying collapse to the late twenty-first century when economic activity overwhelmed improvements in efficiency and pollution control.

The book was subjected to severe criticism and misrepresentation by conventional economists. The principal critiques were based on the false claim that the study predicted that resources would be depleted and the world system would collapse by year 2000. However, in 2008, Graham Turner, a scientist at Australia's national research organisation, the Commonwealth Scientific and Industrial Research Organisation (CSIRO), compared the business-as-usual scenario of the Limits to Growth study with the observed historical data for the period 1970–2000. He found a close match between the model and observation for almost all the outputs reported.[13] An update published in 2017 showed that the world was still following the 'business-as-usual' scenario towards the likely collapse of civilisation.[14]

A recent study by Thomas Cernev considers four scenarios, two of which have high Global Catastrophic Risk (GCR) events[15] (e.g. climate disaster; nuclear war; pandemic) and two with low GCR events. Within each of these categories, there are two scenarios: planetary boundaries have/have not been extensively crossed. Each of the four scenarios represents a possible future world. Considering the current status of planetary boundaries and the likely magnitudes of the risks, Cernev concludes, "*it is evident that in the absence of ambitious policy and near global adoption and successful implementation, the world continually tends towards the Global Collapse scenario.*"[16] Cernev's study was prepared as a Contributing Paper to the 2022 Global Assessment Report on Disaster Risk Reduction of the UN Office of Disaster Risk Reduction,[17] but is not cited in the UN report. It has been suggested that the UN Report has been watered down.[18] Nevertheless, its statement, quoted at the beginning of the chapter, gives a concise, powerful summary of the threats to the environment.

Scenario studies have been complemented by the first comprehensive global assessment of nature that covered the history of humanity's interactions with nature, focusing on the past 50 years, and how these might change in the future. The results were distilled and updated into a major report by an interdisciplinary team of 28 authors from around the world and published in the leading international journal, *Science*. In summary, the team found that:

> The fabric of life on which we all depend—nature and its contributions to people—is unravelling rapidly. Despite the severity of the threats and lack of enough progress in tackling them to date, opportunities exist to change

future trajectories through transformative action. Such action must begin immediately, however, and address the root economic, social, and technological causes of nature's deterioration.[19]

Following *The Limits to Growth* studies, many computer models that combine economic growth, population growth and environmental impacts, including climate change, have been published. Based on these and other studies, Thomas Wiedmann and colleagues published a scientists' warning on affluence.[20] They found that

> Any transition towards sustainability can only be effective if far-reaching lifestyle changes complement technological advancements. However, existing societies, economies and cultures incite consumption expansion and the structural imperative for growth in competitive market economies inhibits necessary societal change.

This means that the people of planet Earth and their governments are faced with a choice of two principal pathways into the future (see Fig. 2.3). One leads to overshoot and collapse; the other leads potentially to an ecologically sustainable future. But a healthy environment is only

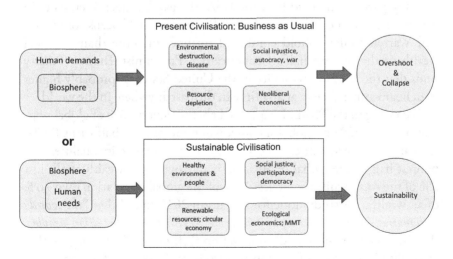

Fig. 2.3 Pathways to two different futures

half the sustainability challenge. We must also work towards social justice, human health and peace. Figure 2.3 illustrates both the social and environmental implications of the scenarios.

2.2   Human Society in Distress

Despite the spread of education and the growth of the middle class, much of the world's human population still suffers from a wide range of crippling afflictions: poverty, economic exploitation, prejudice, war, violent crime, slavery and huge inequalities within and between nations. The inequalities are in wealth, income, resources, health, food, education, housing, legal/justice procedures, personal security and political power.

To cover all these social issues would take several books. Instead, we offer here an indicative sample focusing on wealth and income, first in the world's richest country and then in the whole world.

Inequalities in Income and Wealth

In recent decades, the gap between the rich and poor has been increasing in many parts of the world. In the USA, the gap has been increasing in wealth, income and life expectancy.[21] Three men—Jeff Bezos, Bill Gates and Warren Buffett—hold combined fortunes of more than the total wealth of the poorest half of Americans.[22] Economist Thomas Piketty finds that the level of inequality in the United States is "probably higher than in any other society at any time in the past anywhere in the world."[23]

According to the World Inequality Lab, the richest 10% of global population takes 52% of global income, while the poorest half earns 8.5%. Global wealth inequalities are even greater than income inequalities: the poorest half of the global population barely owns any wealth at all, just 2% of the total, while the richest 10% of the global population own 76% of all wealth.[24] A study done for Oxfam in 2019 found that *"the world's billionaires, only 2,153 people, had more wealth than 4.6 billion people"*, that is, over half the world's population.[25] Thomas Piketty argues that inequality is a feature of capitalism and, as an increasing fraction of the very rich has obtained their wealth (and hence political influence) by inheritance, they are collectively becoming an oligarchy that is a threat to democracy.[26]

Conventional economists argue that, even though the gap between rich and poor is increasing, economic growth increases income of the poor; in other words, economic growth reduces *absolute* poverty, even though *relative* poverty may be increasing. A counter-argument is that the rich have much more influence on political decision-making than the poor and so, as the rich become richer, public facilities that are more important to the poor—public transport, free public education, free or low-cost medical care, low-cost child care, social housing, public parks—are reduced or eliminated. In addition, as the rich become richer, gentrification of low-income areas in inner city suburbs forces the poor out to the urban fringe, where public transport and other facilities are inadequate. In some rich countries, tax incentives encourage middle- and upper-income earners to invest in property in addition to their homes, pushing up house prices for low-income earners. "*The rich get richer with ever more houses; the poor get rent stress or homelessness*".[27] So, it's not obvious that the poor in rich countries necessarily become better off from a small increase in their income derived from economic growth while the rich become much richer. It all depends on the *distribution* of income, wealth and political power about which GDP tells us nothing (see Chap. 7, Sect. 7.2).

In the international sphere, responses to some of these inequalities and injustices are undertaken by United Nations agencies and community-based non-government organisations (NGOs), such as those listed in Table 2.1. The UN Security Council is not listed because it is inhibited

Table 2.1 A selection of international organisations working to improve human society

UN Agencies	NGOs
UN Children's Fund	Amnesty International
UN Development Programme	Human Rights Watch
UN Women	International Campaign to Abolish Nuclear Weapons
World Bank	Médecins Sans Frontières
World Food Programme	Oxfam
World Health Organization	Save the Children

from exercising its main responsibility for the maintenance of peace and security by the use of the veto. Both the UN agencies and the NGOs are limited by insufficient funding compared with the magnitudes of their tasks. Further information about the issues addressed by these organisations is available on their websites and magazines such as *New Internationalist*.

The key point about these afflictions is that almost all of them are not the fault of the victims or of nature, but instead are caused by human institutions: an economic system and political power structures that exploit the environment and people and facilitate inequity, prejudice, violence and war. These problems can be solved, or at least ameliorated, given the political will. As we show in subsequent chapters, the solutions require changes to the economic system, social structures and institutions, and governance. They would be strengthened by an ethical commitment to ecological sustainability and social justice. The main barrier is the power of the minority of people who benefit from the current system.

Who Is Responsible?

Who is responsible for environmental destruction and social injustice? A common answer is that we all are. To some degree, this is true. Each of us uses the energy and material resources, and creates the wastes, that damage the environment. But we are not all equal in unsustainable practices. Each member of the Global North benefits from natural resources extracted from the Global South. In addition, many cheap goods are made by under-paid labour, some of whom are children, living in the faraway lands. Even if a typical citizen of Europe, North America or Australia lives in an all-electric home powered by rooftop solar, travels by public transport and recycles household wastes, their environmental and social impacts, both direct and indirect via their purchased goods and investments, are likely to be much higher than those of a village dweller in India, Bangladesh or Somalia.

Two researchers at the Paris School of Economics, Lucas Chancel and Thomas Piketty, studied the trends in the global inequality of carbon emissions.[28] They found that the top 10% of individual emitters were responsible for 45% of world emissions, while the bottom 50% of individual emitters produced 13% of world emissions. Furthermore, they found that the top 1% of richest Americans, Luxembourgers, Singaporeans and Saudi Arabians are the highest emitters in the world, with annual per capita emissions above 200 tonnes of CO_2-equivalent (CO_2e). Presumably the very rich own private aircraft, luxury motor yachts and multiple palatial homes! The lowest income groups of Honduras, Mozambique, Rwanda and Malawi had emissions of about 0.1 tonnes CO_2e, 2000 times smaller than the super-rich. Similar results were obtained by Jordi Teixidó-Figueras and colleagues, who examined material use and land use as well as carbon emissions [29]: they found that the world's top 10% of income-earners are responsible for between 25% and 43% of environmental impacts, while the bottom 10% of income-earners are responsible for only 3–5% of impacts. Studies in Austria,[30] Australia,[31] the Netherlands[32] and the United Kingdom[33] show that total (direct plus indirect)[34] GHG emissions from individuals or households increase with individual or household income. Typically, emissions from the top decile in countries of the Global North are 3–4 times those of the bottom decile.

The above studies suggest implicitly that substantial environmental improvements could be obtained by policies to limit individual wealth and income. In the context of policy responses, Chancel and Piketty wrote, *"It is then crucial to focus on high individual emitters rather than high emitting countries."*[35] This conclusion is based in part on the fact that there are big gaps in wealth and income within many countries as well as between countries.

While accepting the need to focus on high individual emitters, we also believe that the impacts of rich *countries* should not be ignored, not least because they are best able to take action. Individuals within countries benefit or suffer from the socioeconomic conditions, institutions, infrastructure and natural resources of their countries. As a result, the total (direct plus indirect) greenhouse gas (GHG) emissions per person due to consumption of goods and services by the lowest income quintile of rich countries are generally similar to or greater than the global average emissions of 6.4 tonnes per person. This is confirmed by the specific studies cited above for Austria, Australia, the Netherlands and the UK. Australia is an egregious example (see Box 2.2).

Box 2.2 Case Study: We Must Address the Emissions of Rich Countries as Well as Rich Individuals

Australia has one of the highest average GHG emissions per person in the world, 20 tonnes of carbon dioxide equivalent (CO_2e) according to the usual territorial or production-based accounting.[36] Australia is a rich country, with gross disposable income per household of A$127,000 in 2019–2020, including social assistance benefits, or A$48,846 per person. For the lowest income quintile, gross disposable income in 2019–2020 was A$49,580 per household, including social assistance benefits, or about A$19,000 per person.[37] Although extreme poverty exists in Australia, especially among indigenous people, most of the households in the lowest quintile own a second-hand car and have (rental) housing. Several studies[38] have found that Australia's consumption-based[39] GHG emissions increase with annual per capita income and that the lowest incomes are responsible for 12–17 tonnes of total (direct plus indirect) CO_2e per capita, more than double the global average emissions. Clearly, changes to the economic system are needed in Australia and other rich countries.

Therefore, international agreements and national policies are also needed to constrain the quantity and types of consumption fostered within rich countries. Furthermore, ethical principles would require rich countries to set more rigorous climate targets and policies than the global average, while compensating low-income earners for disadvantage resulting from the policies.

Responsibility for environmental impacts cannot be limited to rich individuals and rich countries. Chapter 6 examines the roles of corporations, especially multinational ones, in capturing the decision-making processes of nation-states. The captured governments of the Global North often assist these corporations in exploiting the Global South.

2.3   Why Does Sustainability Require Social Justice as Well as Environmental Protection?

The role of rich individuals and countries is the first connection between social inequity and environmental impacts. Reducing the income gap between rich and poor, by capping the income of the rich and/or taxing them substantially, while simultaneously improving public facilities (health, education, housing, transport, utilities), would reduce the demand for high incomes, reduce environmental damage and improve social equity.

The second connection arises from ensuring that workers, who lose their jobs as a result of the transition to a more sustainable society and economy, are offered assistance—by government, business or communities—in the form of retraining, relocation, pensions for premature retirement, and opportunities for employment in new, cleaner industries in their local regions. In addition to the ethical reasons for justifying these policies is the practical one of reducing resistance to the transition. We observe that such resistance is understandably substantial in coalmining electorates, where some local politicians are misleading voters that coal mining can and will continue into the long-term future.

The third connection is that the exploitations of the planet and its people are both driven by the dominant economic system comprising neoclassical economic theory and neoliberal economic practice (see Chap. 7). Therefore, changing the economic system would mitigate both forms of exploitation.

The fourth connection is an ethical one: we believe that ecological sustainability should be a key goal for civilisation and that all people should have social sustainability, including better health and living conditions. Therefore, in the sustainable development process, social justice and ecological sustainability should be combined. Self-styled 'renegade' economist, Kate Raworth, has created the concept of 'doughnut economics', which represents both ecological and social sustainability in a single diagram. Her neat summary of our situation is that "*Humanity's 21st century challenge is to meet the needs of all within the needs of the planet*".[40]

The next chapter goes more deeply into sustainability and sustainable development.

Notes

1. United Nations Office for Disaster Risk Reduction (2022). *Global Assessment Report on Disaster Risk Reduction.* https://www.undrr.org/gar2022-our-world-risk

2. World Health Organisation (n.d.). Air quality and health. https://www. who.int/teams/environment-climate-change-and-health/air-quality-and-health/health-impacts/exposure-air-pollution

3. Stockholm Resilience Centre, https://www.stockholmresilience.org/research/research-news/2022-04-26-freshwater-boundary-exceeds-safe-limits.html

4. Lan Wang-Erlandsson et al. (2022). A planetary boundary for green water. *Nature Reviews Earth & Environment,* https://doi.org/10.1038/s43017-022-00287-8

5. Will Steffen et al. (2015). Planetary boundaries: guiding human development on a changing planet. *Science* 347:1259855. https://doi-org.wwwproxy1.library.unsw.edu.au/10.1126/science.1259855

6. Andrea Wulf (2015). *The Invention of Nature: Alexander von Humboldt's new world.* Knopf & John Murray. https://www.andreawulf.com/about-the-invention-of-nature.html

7. IPCC (2022a). *Climate Change 2022: Impacts, adaptation and vulnerability.* Summary for policymakers. https://www.ipcc.ch/report/ar6/wg2/

8. David Armstrong-Mackay et al. (2022). Exceeding 1.5C global warming could trigger multiple climate tipping points. *Science* 377:16611. https://www.science.org/doi/10.1126/science.abn7950

9. NOAA (2022). https://www.climate.gov/news-features/understanding-climate/climate-change-global-temperature

10. Paul Ehrlich & John Holdren (1971) Impact of population growth. *Science* 171:1212–1217; Paul Ehrlich and Anne Ehrlich (2008). Too many people, too much consumption. *Yale Environment 360.* https://e360.yale.edu/features/too_many_people_too_much_consumption

11. Tim Gore (2021). *Carbon Inequality in 2030.* Oxfam. https://www.oxfam.org/en/research/carbon-inequality-2030

12. Donella Meadows et al. (1972). *The Limits to Growth : A report for the Club of Rome's project on the predicament of mankind.* Universal Books.

13. Graham Turner (2008). A comparison of *The Limits to Growth* with 30 years of reality. *Global Environmental Change* 18:397–411.

14. Graham Turner (2014). Is global collapse imminent? MSSI Research Paper No. 4, Melbourne Sustainable Society Institute, University of Melbourne. https://sustainable.unimelb.edu.au/__data/assets/pdf_file/0005/2763500/MSSI-ResearchPaper-4_Turner_2014.pdf

15. A GCR event is defined to be one that results in 10 million fatalities or greater than USD$10 trillion in damages; the damage must be extensive and on a global scale.

16. Thomas Cernev (2022). *Global Catastrophic Risk and Planetary Boundaries: The relationship to global targets and disaster risk reduction.* https://www.undrr.org/publication/global-catastrophic-risk-and-planetary-boundaries-relationship-global-targets-and
17. United Nations Office for Disaster Risk Reduction (2022), *op. cit.*
18. Nafeez Ahmed (2022). UN warns of 'total societal collapse' due to breaching of planetary boundaries. *Byline Times*, 26 May. https://bywire.news/articles/un-warns-of-total-societal-collapse-due-to-breaching-of-planetary-boundaries
19. Sandra Diaz et al. (2019). Pervasive human-driven decline of life on Earth points to the need for transformative change. *Science* 366:1327. This is just a one-page summary; the full review is available free at https://doi.org/10.1126/science.aax3100
20. Thomas Wiedmann et al. (2020). Scientists' warning on affluence. *Nature Communications* 11:3107. https://doi.org/10.1038/s41467-020-16941-y
21. Lola Fadulu (2020). Study shows income gap between rich and poor keeps growing, with deadly effects. *New York Times*, 11 June. https://www.nytimes.com/2019/09/10/us/politics/gao-income-gap-rich-poor.html
22. Inequality.org. https://inequality.org/facts/wealth-inequality/
23. Thomas Piketty (2014). *Capital in the Twenty-First Century.* Harvard University Press.
24. Lucas Chancel et al. (2022). *World Inequality Report 2022.* World Inequality Lab. https://wir2022.wid.world/
25. Max Lawson et al. (2020). *A Time to Care: Unpaid and underpaid care work and the global inequality crisis.* Oxfam. https://oxfamilibrary.openrepository.com/bitstream/handle/10546/620928/bp-time-to-care-inequality-200120-en.pdf
26. Piketty (2014), *op. cit.*
27. Tone Wheeler (2022). https://www.theguardian.com/australia-news/2022/mar/13/we-must-separate-the-idea-of-house-from-home-the-case-for-drastic-action-on-shelter
28. Lucas Chancel & Thomas Piketty (2015). *Carbon and inequality: from Kyoto to Paris.* Paris School of Economics. https://doi.org/10.13140/RG.2.1.3536.0082
29. Jordi Teixidó-Figueras, J. et al. (2016). International inequality of environmental pressures: decomposition and comparative analysis. *Ecological Indicators* 62:163–173.

30. Hendrik Theine et al. (2022). Emissions inequality: disparities in income, expenditure, and the carbon footprint in Austria. *Ecological Economics* 197:107435, Fig. C.3. https://doi.org/10.1016/j.ecolecon.2022.107435

31. Manfred Lenzen et al. (2002). A personal approach to teaching about climate change. *Aust. J. Environ. Education* 18:35–45, Fig. 6; Manfred Lenzen & Robert Cummins (2013). Happiness *versus* the environment—a case study of Australian lifestyles. *Challenges* 4:56–74. http://www.mdpi.com/2078-1547/4/1/56, Fig. 2.

32. Annmarie Kerkhof et al. (2009). Relating the environmental impact of consumption to household expenditures: an input-output analysis. *Ecological Economics* 68:1160–1170. https://doi.org/10.1016/j.ecolecon.2008.08.004

33. Milena Büchs & Sylke Schnepf (2013). Who emits most? Associations between socio-economic factors and UK households' home energy, transport, indirect and total emissions. *Ecological Economics* 90:114–123, Graph 1. https://doi.org/10.1016/j.ecolecon.2013.03.007

34. Direct emissions are those resulting from people directly using energy by for example, operating electrical appliances, a car or a home heater. Indirect emissions are those embodied in products people buy, imported as well as manufactured locally. In many countries of the Global North, indirect GHG emissions are much greater than direct emissions.

35. Chancel & Piketty, *op. cit.*

36. Department of Climate Change, Energy, the Environment and Water (2022). *Quarterly update of Australia's national greenhouse gas inventory: December 2021.* https://www.dcceew.gov.au/climate-change/publications/national-greenhouse-gas-inventory-quarterly-update-december-2021

37. Australian Bureau of Statistics (2021). https://www.abs.gov.au/statistics/economy/national-accounts/australian-national-accounts-distribution-household-income-consumption-and-wealth/latest-release

38. Lenzen et al. (2002), *op. cit*; Lenzen and Cummins (2013), *op. cit*.

39. According to the international greenhouse accounting convention, official measurements are limited to emissions within the territory of the country concerned, known as territorial or production-based emissions. However, these emissions do not take account of the emissions embodied in imports. The discussion in Box 2.2 relating household income to

GHG emissions requires consumption-based household emissions comprising both direct emissions (e.g. from using electricity, petrol and natural gas) and indirect emissions embodied in imported goods purchased by the household. In general, indirect emissions in Global North countries are greater than direct emissions.

40. Kate Raworth (2018). *Doughnut Economics: Seven ways to think like a 21st-century economist.* Cornerstone & Chelsea Green, https://www.kateraworth.com/doughnut/

Reference

Weblinks accessed 26/10/2022.

3

An Ecologically Sustainable, Socially Just Civilisation

Treating the world as if we intended to stay. (Crispin Tickell)[1]

A group of enthusiastic hikers were driving around in unfamiliar country. They didn't have an accurate map and were lost. They stopped their car alongside a farmer who was mending his fence. "Good day", they said, "we would like to climb that mountain we can see in the distance. Can you tell us how to get there?" The farmer thought for a moment before replying: "Well, if I were going to the mountain, I wouldn't start from here and I wouldn't attempt to climb it in this hot weather."

This variation on an old story illustrates the situation of many of us citizens of planet Earth. We wish to travel to a better place, a distant mountain called Sustainability, more precisely, an ecologically sustainable and socially just society, but we are starting from the society and civilisation described in Chap. 2, which is destroying its life support system and lacks social justice. Because many of the threats we are facing are global rather than local, we understand that a sustainable society must become the destination for the whole world. We have called this global sustainable society 'the Sustainable Civilisation' and the pathway or pathways to reach it, 'sustainable development' (SD), which we use as a shortened

© The Author(s), under exclusive license to Springer Nature Singapore Pte Ltd. 2023
M. Diesendorf, R. Taylor, *The Path to a Sustainable Civilisation*,
https://doi.org/10.1007/978-981-99-0663-5_3

version of 'ecologically sustainable and socially just development'. Although the goal, sustainability, and the transition process, sustainable development, are often confused in the literature, they must be distinguished in a thoughtful analysis.

The aim of this chapter is to clarify these concepts and introduce some of the characteristics needed to achieve the Sustainable Civilisation. The following chapters explore these characteristics in more detail.

3.1   Sustainability and Sustainable Development

Of all words in the environmental vocabulary, there is probably none more abused than 'sustainable'. The Concise Oxford Dictionary of Current English has eight definitions of 'sustain': from these, we wish to apply to human civilisation and the biosphere, 'enable to last out'; 'keep from failing'; and 'keep going continuously'. For the term 'sustainable development', literally hundreds of different definitions, that emphasise different aspects of the concept, have been published in the scholarly literature. One of the earliest and best-known was published in 1987 by the World Commission on Environment and Development: "*Sustainable development is development that meets the needs of the present, without compromising the ability of future generations to meet their own needs*".[2] While offering the important perspective of intergenerational equity, this definition raises several questions: What are genuine needs as opposed to desires? Does development necessarily entail economic growth? What about intragenerational (social) equity? Would an ecocentric concept be preferable to an anthropocentric (human-centred) one?

The lack of precision in definitions of sustainability and SD has enabled their misuse by vested interests to justify continuing their environmental damaging activities. For example, in North America, the so-called Sustainable Forestry Initiative has been rightly criticised by many environmental organisations—for example, Sierra Club, Greenpeace, Natural Resources Defense Council, Rainforest Action Network—for approving clear-cutting of native forests and their conversion to plantations, the destruction of biodiversity, excessive spraying with toxic pesticides and the failure to require the consent of indigenous peoples.[3] Such examples have led to critiques of the term sustainable development[4] and demands that it be discarded. Nevertheless, sustainability and SD are important concepts that, by their very nature, cannot be defined in a precise manner. They are contestable concepts, like justice, freedom and democracy, whose meanings emerge from continuing discussion, debate and social practice. Like justice, freedom and democracy, sustainability and SD are worth struggling for and should not be dismissed on the shaky grounds that they cannot be defined precisely like the standard metre and so can be appropriated and misused by vested interests.

If we are serious about environmental protection and our own survival, we need 'strong' definitions of sustainability and SD that do not involve trade-offs between the environment and the economy. Such definitions recognise that we humans are totally dependent upon the natural environment for the air we breathe, a stable climate, the food we ingest and digest, and the natural bio-geo-chemical cycles, especially the carbon dioxide (CO_2), phosphorus, water, oxygen and nitrogen cycles.[5] When big business introduces its concept of SD, it often uses the word 'balance' as a code for trade-offs between the environmental and the economy. It may wish to trade-off a forest against a coal mine or a new suburb, a coral reef against a massive tourist development with chlorinated 'infinity pools', a wilderness valley against a motorway. On a finite planet, continuing trade-offs will inevitably result in the destruction of our life-support system, as surely as a massive comet colliding with the Earth. They are not 'sustainable' in the strong sense that the planet and its peoples need.

'Development' is also a problematic term because it is often confused with economic growth. In this book, in the context of sustainable development, 'development' means social and economic improvement in a broad sense, with emphasis on qualitative improvement in human wellbeing or unfolding of human potential. It may or may not involve economic growth as measured quantitatively by an increase in gross domestic product (GDP).

For the process of transitioning to the Sustainable Civilisation, we interpret the term 'sustainable development' as 'ecologically sustainable and socially just development', with development interpreted as in the previous paragraph.[6]

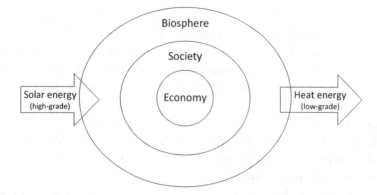

Fig. 3.1 The relationship between the biosphere, society and economy in a sustainable society

An example of a strong definition of SD that avoids trade-offs is, "*Sustainable development comprises types of economic and social development that protect and restore the natural environmental and social justice*".[7] In other words, economic and social developments are constrained by the environmental limits of a finite Earth. We must recognise the primary role of our life support system, that human society is embedded within the biosphere and must be compatible with it, and that the economy is a human construct that must be compatible with society, as illustrated in Fig. 3.1. Neoclassical economics reverses the order, giving primacy to economics, as discussed in more detail in Chap. 7. Note that the whole system is driven by the input of high-grade energy from the Sun, while the Earth emits low-grade energy in the form of heat at low temperature.[8]

Crispin Tickell's concise definition of sustainable development, quoted at the beginning of the chapter, could also be interpreted as a strong definition, at least in terms of environmental impact.

Indigenous people were acutely aware of the need to care for country and conserve its natural resources in order to survive and thrive. We can learn much from them (see Box 3.1).

Box 3.1 Sustainability of Indigenous People

Around fifty thousand years ago, when Neanderthals were the dominant people in Europe, modern (*Homo sapiens*) humans were already living in the land we now call Australia.[9] The first Australians lived primarily by gathering food, supplemented by hunting and, for some tribes, agriculture. Despite the tragic impact of settler colonisation on the population and cultures of the Australian Aborigines from 1788 onwards, some tribal groups are still living today on their traditional lands, over which they never ceded sovereignty.

Long before the arrival of Europeans, the Aborigines modified the land with the controlled use of fire to clear the underbrush and make travel easier, to hunt large and small game, to increase the abundance of certain types of plant foods, and to manage wild fires.[10] Scientific debate continues about whether the Aborigines, or climate change, was responsible for the extinction of the megafauna, the giant reptiles, marsupials and birds that once roamed the land.[11] Nevertheless, their 50,000 years of continuous occupation must be recognised as an achievement of a people who advanced a long way towards sustainability, especially when compared with the serious threat of the future collapse of global industrial society in the twenty-first century.

What can we learn from the Aborigines about sustainable living? Although the present huge human population of this planet cannot return to a gatherer-hunter lifestyle, we can learn from the fact that Aboriginal people developed close, nurturing relationships with the land and its other living creatures. Each tribe identified with its local 'country' and learned the details of its landforms, waterways and waterholes, plants and animals. This knowledge was taught by the elders to the younger generations. It was embodied in spiritual and cultural practices passed down from the 'dreamtime' and in rock engravings and paintings which predate the cave art in Europe. Each person had a totem or dreaming, a special, caring relationship with a natural species, usually an animal or plant; clans and other groups within the tribe also had totems.[12] Special areas, designated 'story places' which humans never entered, were protected habitat for fauna necessary for the survival of the tribe.[13] When the Aborigines gathered plant foods, they took care not to uproot the whole plant. They were, of necessity, conservationists. This contrasts with industrial society, a society dominated by takers.

The inclusion of social justice/equity in SD is first and foremost an ethical requirement that's consistent with the Universal Declaration of Human Rights. These rights include the right to take part in the government of one's country, directly or through freely chosen representatives

(Article 21), to social security (Article 22), to work for just remuneration (Article 23), to a standard of living adequate for health and wellbeing (Article 25) and education that's free, at least in the elementary and fundamental stages (Article 26).[14]

As discussed in Chap. 2, Sect. 2.3, social justice is vital for practical reasons as well as ethical. Without the inclusion and active participation of low-income countries and low-income people within both rich and poor countries, global environmental protection would become almost impossible. Some of the most serious problems facing human civilisation and the survival of the planet are global in nature, for example, climate change, biodiversity loss, poverty and a rapidly mutating virus that knows no borders. Cooperation is needed to solve them.

3.2   Sustainable Development Goals

In 2015, the United Nations Sustainable Development Summit adopted 17 Sustainable Development Goals (SDGs) as part of the United Nations 2030 Agenda for Sustainable Development.[15] The goals span environment (including climate), health, education, poverty, inequalities within and among countries, gender equality, peace and justice.

While the goals are a good start, there is a fundamental contradiction between Goal 8 and Goal 12. Goal 8 aims "*to promote sustained, inclusive and sustainable economic growth, full and productive employment and decent work for all*", while Goal 12 aims "*to ensure sustainable consumption and production patterns*". The problem is that economic growth is equivalent to increasing consumption which drives increasing environmental impacts. The unrealistic assumption underlying Goal 8 is that economic growth can somehow be dematerialised, that is, decoupled from physical growth, leading to the notion of 'green growth'. The European Green Deal makes the same unjustified assumption as Goal 8, specifying 'economic growth decoupled from resource use'. The American Green New Deal simply assumes continuing economic growth.

But so far economic growth has been closely coupled to growth in the use of energy (see Chap. 4) and materials (see Chap. 5), and to growth in population.[16] Each of these physical growths has environmental impacts. Decoupling has only been observed in particular countries for short periods.[17] While green growth will generally have lower resource use and environmental impacts than business-as-usual growth, even labour-intensive service industries have significant indirect environmental impacts: school teaching requires school buildings and transport; home care for the aged requires equipment and transport; tourism involves travel, accommodation and other facilities; office work requires office equipment and, in many cases, office buildings. Although the direct environmental impacts of most renewable energy technologies are very low,[18] they have the same kinds of *indirect* environmental impacts as any other energy source that turns the wheels of industry and transportation.

Neoclassical economic ideology[19] fosters endless economic growth, resulting in increasing environmental damage. It has failed to solve the problem of poverty—indeed, the gap between the rich and poor is increasing globally.[20, 21] The continuing growth in consumption, at least by the rich countries, is unsustainable. Therefore, the transition to environmental sustainability and social justice must be accompanied by strategies that lead to a steady-state economy (SSE) with consumption reduced to a level that's compatible with sustainability. This is discussed in more detail in Sect. 3.3 and Chap. 7.

Our suggested revised SDG Goal 8 is "*to promote sustained, inclusive and sustainable quality of life, full and productive employment and decent work for all*". With this revision and the closure of a gap in the SDG goals regarding population (see below), the SDGs would summarise the properties of a sustainable society that could lead to the Sustainable Civilisation (see Box 3.2).

Box 3.2 Specific Environmental and Resource Goals of the Sustainable Civilisation

- returning to a stable climate with global heating reduced to a level that's as far as possible below 2 °C above the pre-industrial level
- reducing the rate of biodiversity loss to the pre-industrial level or less
- restoring the quality of our soils
- restoring our forests
- substantially reducing the pollution of air, water and soils
- maintaining the flows of rivers and reducing the extraction of ground water
- more generally, reducing the use of non-renewable resources as far as possible and reducing the rate of use of renewable resources to below the rate of natural replenishment

Biologist and naturalist Edward O. Wilson has argued that, to protect our endangered biosphere, piecemeal policies are inadequate in the face of current degree of destruction, and therefore half the surface of the Earth should be dedicated to nature.[22] Two small steps in this direction would be to strengthen the protections of World Heritage Convention[23] and to extend the Antarctic Treaty[24] to more parties and for an unlimited time.

At very least, the social equity goals can be interpreted to require that all people should have equal opportunity in:

- basic housing, health care and education
- access to renewable and non-renewable resources
- in particular, access to energy that's clean, reliable and affordable
- access to sufficient nutritious food and clean drinking water
- personal security, that is, freedom from physical and mental violence, and from arbitrary arrest and imprisonment without due process; the right to a fair trial
- community participation in national, state and local government decision-making
- peaceful coexistence, not war, between and within nations
- stable population. (Surprisingly, this has been omitted from the SDGs, although population is one of the drivers of adverse environmental and social impacts.)

Manfred Max-Neef, known as 'the barefoot economist', offers a framework that treats human needs in a system that distinguishes fundamental needs from the means of satisfying them. He identifies nine types of fundamental needs: subsistence, protection, affection, understanding, participation, idleness, creation, identity and freedom.[25]

The emphasis here on satisfying fundamental human needs is deliberately very different from the neoliberal economics assumption that all needs are valid and the market should decide which are delivered and in what quantities. This extreme market-driven approach is inconsistent with SD, as discussed in the next section.

3.3   The Sustainable Civilisation

So far, our ecologically sustainable and socially just goal for global society, the Sustainable Civilisation, appears similar to the Green New Deal proposed for the USA[26] and the European Green Deal.[27] However, there is an important difference: the two Green Deals have been developed within the framework of conventional (neoclassical) economics, which assumes that economic growth can continue, while the Sustainable Civilisation has limits to growth, at least to physical growth.

The conceptual framework of neoclassical economics, which its proponents consider to be a discipline and critics consider to be an ideology or quasi-religion, is based in part on the separation of the economy from the natural environment. It treats the environment as an infinite resource base and an infinite waste bin (see Chap. 7, Sect. 7.1 for more detail).

It extracts from our planet both renewable resources, such as water, timber and fish, and non-renewable resources, such as minerals and fossil fuels.[28] If the rate of extraction of renewable resources is greater than the rate of natural replenishment, then the stock of renewables declines. All extraction of non-renewable resources reduces the available stock. The resources extracted from the environment are transformed into 'goods' and services, with production driven by consumption, and consumption driven by the ideology of consumerism supported by advertising.[29] The wastes, including the original products at the end of their lifetimes, are dumped into our air, waters, soils and landfill.

By contrast, the interdisciplinary field of ecological economics rejects the neoclassical economics conceptual framework and neoliberal economics practice (see Chap. 7, Sect. 7.2). It recognises that, on a finite planet, continuing growth in resource use and environmental pollution decimates the biosphere, the basis of life on Earth. This understanding is based on science that shows that the existing human civilisation is exceeding the physical limits of our planet (see Chap. 2). This is very different from the unrealistic assumptions underlying conventional economics.

By drawing upon ecological economics and conservation biology, we find that the Sustainable Civilisation must have physical limits as well as the cleaner technologies and industries envisaged in the Green Deals.

Specifically, the Sustainable Civilisation must have limits on the use of energy, materials and land, and on populations, both human and certain farmed animals. In other words, it works towards a steady-state economy, an economy with no physical growth.

But the global economy operating at current levels of throughput or consumption is not ecologically sustainable, as discussed in Chap. 2. Therefore, the rates of use of energy and materials must be reduced to levels that are ecologically sustainable. It is necessary, but not sufficient, to transform fossil-fuelled energy entirely to renewables. Even renewable energy at the current rate of consumption would turn the wheels of transport and industry at rates that damage the environment irreversibly. Therefore, global energy consumption must be reduced while transitioning to renewables (see Chap. 4). Limiting the rate of materials consumption requires, as far as possible, a circular economy, an economy based on the principles of reduce, reuse, recycle, recover, redesign and remanufacture. The key phrase here is 'as far as possible', because a completely circular economy is impossible (see Chap. 5, Sect. 5.4).

In summary, the Sustainable Civilisation will need a steady-state economic system defined as *an economy with constant stocks of people and products, maintained with low levels of energy and materials throughput that are ecologically sustainable.* This is very similar to the definition given by in 1977 by one of the founders and leaders of ecological economics, Herman Daly.[30] To transition to this steady-state economy, the sustainable development process must involve *degrowth* in physical terms, at least for the rich countries. Degrowth is the *planned reduction in the use of energy, materials and land, together with stabilisation of the population, with the aim of transitioning to an ecologically sustainable, socially just society.*

To bypass the debate as to whether absolute decoupling between economic growth and physical growth is possible, we have developed the concept of the Sustainable Civilisation in terms of green growth with *physical* constraints on global resource use. If this places constraints on economic activity, then so be it. Anyway, Gross Domestic Product is a poor measure of human wellbeing. Better measures exist.[31]

While a steady-state economy has constraints on growth of the physical economy, this does not have to result in a society that ceases to develop. In the words of Herman Daly:

> The culture, genetic inheritance, goodness, ethical codes, and so forth embodied in human beings are not held constant. Likewise, the embodied technology, the design, and the product mix of the aggregate total stock of artifacts are not held constant. Nor is the current distribution of artifacts among the population taken as constant. Not only is quality free to evolve, but its development is positively encouraged in certain directions.[32]

Daly distinguishes between development, meaning qualitative change, and growth, meaning quantitative change, and writes, *"a steady-state economy develops but does not grow"*.

While the global steady-state economy ceases to grow, there will be substantial growth and degrowth within it. The Sustainable Civilisation will not stagnate. The phase-out of fossil fuels will be offset by growth in renewable energy, energy efficiency and energy conservation (Chap. 4). Decline in the number of private cars and the land area devoted to roads and parking will be accompanied by growth in public transport, cycling and walking. More resources will flow into public health, public education, public/social housing, the public service and national parks—less into their private equivalents, financial institutions and advertising.

The Sustainable Civilisation is our vision of a diversity of sustainable societies on a global scale, united by the common principles of ecological sustainability and social justice. It involves all people and all countries through global communications, global trade, strengthened international institutions both at government-to-government and at people-to-people spheres, and global cooperation in responding to existential threats. To be accepted by the majority of the world's people, the transition to the Sustainable Civilisation must reduce the inequalities in access to natural resources, and in political and economic power between countries, regions and communities. To achieve this, it must overcome the resistance of the rich and powerful.

3.4   Sustainable Development of the Global South

To understand the situation of countries and regions where the majority of the population is poor and powerless, called the Global South,[33] we must come to grips with colonialism and neo-colonialism.

Great civilisations ruled for centuries in Mesoamerica,[34] Africa[35] and East Asia[36]. Small communities of indigenous people lived for millennia in all the habitable continents. The invading European colonial powers decimated these civilisations and communities by force of arms, the diseases they brought, slavery, drugs such as alcohol and opium, land grabbing, and robbery of natural resources. They reinforced their control by racism, exclusion of the local people from their traditional lands, cultural assimilation, the removal of children from their parents for 're-education', and the Christian religion.[37] Colonisation also created a mindset among the colonised that "*the colonial power had set the nation on the right path and the only problem* [after decolonisation] *was how to replace 'them' by 'us'.*"[38]

Between 1945 and 1960, many countries obtained their independence from colonial powers. Decolonisation occurred by peaceful negotiations, violent revolts and nonviolent resistance.[39] It created, momentarily, the illusion of freedom and independence among the Global South. However, it was followed immediately by neo-colonisation: the imposition by the West of the capitalist economic system with investments in resource extraction by multinational corporations that cause economic, social and environmental damage; huge loans that often result in foreign debt that cannot be repaid; sponsored military coups and civil wars; and 'development' aid tied to the donors' industries. According to Kwame Nkrumah, the first Prime Minister and President of Ghana, "*Neo-colonialism is also the worst form of imperialism. For those who practise it, it means power without responsibility and for those who suffer from it, it means exploitation without redress.*"[40] Neo-colonialism was established through cooperation between the neo-colonisers and the ruling class of the neo-colonised who are generally rich.[41]

Wolfgang Sachs describes the original 'development' aid programs as having the hidden agenda of "*the Westernization of the world*", and the creation of "*a monoculture [that] has eroded viable alternatives to the industrial, growth-oriented society and dangerously crippled humankind's capacity to meet an increasingly different future with creative responses*".[42] Similarly, the radical scientist Vandana Shiva sees the concept of 'development', which should mean the improved wellbeing for everyone, being reshaped into 'westernisation' of human needs:

> Concepts and categories about economic development and natural resource utilisation that had emerged in the specific context of industrialisation and capitalist growth in a centre of colonial power, were raised to the level of universal assumptions and applicability in the entirely different context of basic needs satisfaction for the people of the newly independent Third World countries.[43]

She quotes Rosa Luxemburg that "*colonialism is a constant necessary condition for capitalist growth: without colonies, capital accumulation would grind to a halt.*" Thus, development was reduced to a process of colonisation and, Shiva argues, was based on "*the exploitation and exclusion of women (of the west and non-west), on the exploitation and degradation of nature, and on the exploitation and erosion of other cultures.*" In India, rural women, who are still embedded in nature, have led struggles to conserve forest, land and water.[44]

Over several decades, the concept of development has evolved and broadened to some extent, thanks to community-based development NGOs in both donor and recipient countries and to some academics who are critical of a system that fosters control by, and dependence upon, the donor or multinational corporations. Nowadays, many non-profit NGOs view the purpose of development as to facilitate the development of self-reliant communities in the recipient countries, the opposite of neo-colonialism. The United Nations Sustainable Development Goals bring together ecologically sustainable development with social justice in international aid. The most successful aid projects are often people-to-people at the community level rather than national.

However, some government-to-government projects are still designed to create trade instead of aid, that is, to provide sales for business and industry based in the donor country, and to exercise control over the recipient countries. For this level of aid, bribery and corruption are common. Of even greater concern is that this kind of development still tends to emphasise economic growth, as measured by GDP, rather than community development and social justice. The trickle-down theory of development is still influential, despite its many failures. The pathway to becoming a 'developed' country is still seen by donors—such as the former colonial powers, multinational corporations and the World Bank—and recipient governments as primarily by exporting the country's natural resources—minerals, forests, fish and arable land—supplemented by tourism. From the viewpoint of the 'less developed' country, the purpose of these exports is to earn foreign currency to try to pay off the nation's debt to donors and to pay for imported Western technology, manufactured goods and food products purchased by the country's rich elite. Thus, one form of colonialism has been replaced by another (see also Chap. 6, Sect. 6.3).

Therefore, genuine sustainable development of the Global South must involve a struggle for freedom from both types of colonialism. The victory of socialist candidate Pedro Castillo in the 2021 Peruvian general election was a possible step towards defeating neo-colonialism in that country. However, an opposition-led alliance controls Peru's Congress. Another approach towards eroding neo-colonialism is the formation of multinational alliances of countries of the Global South. The 2014 Declaration of Havana by 33 national governments of Latin America and the Caribbean rejected the Pan-American Project, a proposed integration of the region directed by the United States.[45]

3.5   What Kind of Sustainable Society?

Many kinds of sustainable society have been proposed. A thought-provoking article by Thomas Wiedmann and colleagues classifies them into three main groups: radical, reformist and green growth.[46] In designing the Sustainable Civilisation, we rejected green growth because it is ecologically unsustainable, as discussed above. The Sustainable Civilisation has both reformist and radical elements. It is reformist in that it's a green industrial society with nation-states and markets, although the latter are firmly guided by governments, institutions and social sanctions. It is radical in that it rejects neoliberal economics and undertakes planned degrowth to a steady-state economy (see Chap. 7), reduces the power of large corporations and other elites, and creates new institutions of governance and other social organisations (see Chap. 6).

Some people recommend a return to a simple, grassroots, self-sufficient, pre-industrial society, with local communities growing their own food, ploughing their fields with draught animals, making their own tools, heating their homes with firewood, and travelling in horse-drawn carriages. While recognising that such local communities can exist within the Sustainable Civilisation and indeed would enhance it, we have not chosen this scenario for the national and global scales for several reasons.

Earth's human population has just passed eight billion and the future mothers of the next generation have already been born. Time is of the essence for transitioning to a sustainable society, but there is no way, except by the collapse of existing civilisation—through nuclear war or pandemic or overshoot—to reduce the global population *rapidly* to a level that could possibly be compatible with a pre-industrial scenario. Such a civilisation would have to be much more dispersed than the present highly urbanised civilisation. Its impact on soils, waterways, forests and biodiversity would be even greater than that of the present civilisation, unless its population were far below the present, possibly less than one billion.

If the pre-industrial society is reached by the collapse of the existing civilisation, the result will, almost certainly, be chaotic. Some local communities can and do plan for societal collapse, for example the heavily armed 'survivalist' communities in the USA, but these are not credible models of a sustainable society. Collapse is a disaster scenario and so it would be difficult to plan for it to transition to an ordered pre-industrial society on national or international scales. At best, one could hope for a slow recovery to rebuild a civilisation, perhaps similar to that described in the post-apocalyptic novel, *A Canticle for Leibowitz*.[47]

In transitioning to the Sustainable Civilisation, we envisage that global population is gradually, non-violently reduced. If the new society is to thrive and if the rich countries choose a green industrial society, then even with half or one-quarter of the present global population, it will require solar panels, railways, bicycles, tractors, efficient electric appliances, ultrasound and X-ray medical technologies, vaccines and antibiotics. These cannot be manufactured by small, local communities, which will still be partially dependent upon a national and international industrial society for steel, aluminium, electronics, chemicals, and so on. Nevertheless, our conception of the Sustainable Civilisation does not preclude some local communities from choosing a simple, self-sufficient, pre-industrial lifestyle and creating their 'earth gardens'. It's just not possible for a whole country or the whole world, if we want the benefits of a green industrial society.

Some members of the Global South are exploring alternatives to western forms of 'development'. Ecuador and Bolivia are building on the worldview of the Indigenous peoples of the Andes to incorporate *sumac kawsay* (in Kichwa) or *buen vivir* (in Spanish) into their approaches. It's a pathway that's community centred, ecologically sustainable and culturally sensitive. It is also influenced by western critiques of capitalism and it doesn't require a return to the Indigenous past.[48]

Swaraj is an old Indian philosophy of self-determination and collective decision-making. Based in part on swaraj, *ecological swaraj*, of the twentieth century onwards, is:

> a framework that respects the limits of the Earth and the rights of other species, while pursuing the core values of social justice and equity. With its strong democratic and egalitarian impulse, it seeks to empower every person to be a part of decision making, and its holistic vision of human well-being encompasses physical, material, socio-cultural, intellectual, and spiritual dimensions.[49]

Ecological swaraj is a bottom-up approach based on collectives and communities. Like buen vivir, it is an evolving worldview.

3.6   Concluding Remarks

We have chosen the Sustainable Civilisation, because it's a scenario of transition without collapse, and hence it's a scenario of hope. Given appropriate institutions, the people can plan and implement it on sub-national, national and international scales. The rich, over-developed countries can start by ending growth in the consumption of energy, materials and land, and by introducing policies to stabilise their populations, while simultaneously closing their highly polluting industries, replacing them with green technologies and shifting, as far as possible, to a circular economy. The rich countries will have to assist financially the low-income countries of the Global South, while allowing the latter to choose their own pathways to a sustainable future, a delicate balance. While the interests of big business currently stand firmly in the way of these strategies, as climate change impacts increase at a frightening pace the alternatives will become increasingly sought after. We believe it may still be possible to transition to the Sustainable Civilisation, but each additional year of delay makes it more difficult.

Notes

1. Crispin Tickell (2002) Human responsibility in the global environment. *Asian Affairs* 33(2):206–215. https://doi.org/10.1080/714041469
2. World Commission on Environment & Development (1987). *Our Common Future*. The Brundtland report. Oxford University Press.
3. Stand.earth website. https://www.stand.earth/page/forest-conservation/primary-and-intact-forests/environmental-leaders-critique-sfi
4. For example, Sharon Beder (1993). *The Nature of Sustainable Development*. Scribe.
5. Haydn Washington (2013). *Human Dependence on Nature*. Earthscan from Routledge.
6. This was the original concept as defined in the 1990–91 Australian Ecologically Sustainable Development process, but subsequently the Australian government redefined it to omit social justice principle—see Clive Hamilton & David Throsby (eds.) (1997). *The ESD Process:*

Evaluating a policy experiment. Academy of the Social Sciences in Australia and Graduate Program in Public Policy, Australian National University, ISBN 0646365231.

7. Mark Diesendorf (2000). Sustainability and sustainable development. In: Dexter Dunphy, Jodie Benveniste, Andrew Griffiths & Philip Sutton (eds.) *Sustainability: The corporate challenge of the 21st century.* Allen & Unwin.

8. Solar energy is deemed 'high-grade' because it is emitted at a temperature of about 6000°K and hence, according to basic physics, has a range of short (optical) wavelengths that carry high energy, while heat energy is deemed 'low-grade' because it is emitted from the Earth with an average temperature of 287°K (14°C) and hence has longer (infrared) wavelengths that carry low energy.

9. Alan Cooper et al. (2018). When did aboriginal people first arrive in Australia? *The Conversation,* 7 August. https://theconversation.com/when-did-aboriginal-people-first-arrive-in-australia-100830; James O'Connell et al. (2018). When did *Homo sapiens* first reach Southeast Asia and Sahul? *PNAS* 115:8482–8490. https://www.pnas.org/doi/epdf/10.1073/pnas.1808385115

10. J.L. Kohen (1996). Aboriginal use of fire in southeastern Australia. *Proc. Linnean Soc. NSW* 116:19–26; Christopher Gillies (2017). Traditional Aboriginal burning in modern day land management. *Landcare Australia.* https://landcareaustralia.org.au/project/traditional-aboriginal-burning-modern-day-land-management; Miki Perkins (2020). What is cultural burning? *Sydney Morning Herald.* https://www.smh.com.au/environment/climate-change/what-is-cultural-burning-20200228-p545e2.html

11. Michael Westaway et al. (2017). Aboriginal Australians coexisted with the megafauna for at least 17,000 years. *The Conversation,* 12 January. https://theconversation.com/aboriginal-australians-co-existed-with-the-megafauna-for-at-least-17-000-years-70589; Frédérik Saltré et al. (2019). Did people or climate kill off the megafauna? Actually, it was both. *The Conversation,* 4 December. https://theconversation.com/did-people-or-climate-kill-off-the-megafauna-actually-it-was-both-127803

12. M.H. Monroe (2011). Aboriginal totemism. https://austhrutime.com/aboriginal_totemism.htm

13. Tim Flannery (1994). *The Future Eaters.* Reed Books, p. 290.

14. United Nations website. https://www.un.org/en/about-us/universal-declaration-of-human-rights

15. United Nations website. *The 17 Goals.* https://sdgs.un.org/goals

16. Economics Discussion. https://www.economicsdiscussion.net/economic-development/population-growth-and-economic-development-2/26308

17. Corinne Le Quéré et al. (2019). Drivers of declining CO_2 emissions in 18 developed economies. *Nature Climate Change, 9,* 213–217; Helmut Haberl et al. (2020). A systematic review of the evidence on decoupling of GDP, resource use and GHG emissions, part II: synthesizing the insights. *Environ. Res. Lett. 15,* 065003. https://doi.org/10.1088/1748-9326/ab842a

18. The exceptions are hydroelectricity based on large dams and bioenergy based on unsustainable sources and conversion methods.

19. Investopedia. https://www.investopedia.com/terms/n/neoliberalism.asp

20. Thomas Wiedmann et al. (2020) Scientists' warning on affluence. *Nature Communications,* 11, 3107, https://doi.org/10.1038/s41467-020-16941-y

21. Benedikt Bruckner et al. (2022). Impacts of poverty alleviation on national and global carbon emissions. *Nature Sustainability,* https://doi.org/10.1038/s41893-021-00842-z

22. Edward O. Wilson (2016). *Half-Earth: Our planet's fight for life.* Liveright.

23. UNESCO. *World Heritage.* https://whc.unesco.org/en/about/

24. Secretariat of the Antarctic Treaty. https://www.ats.aq/index_e.html

25. Manfred Max-Neef (1991). *Human Scale Development: Conception, application and further reflections.* Apex Press. http://www.wtf.tw/ref/max-neef.pdf

26. Green Party US. *Green New Deal.* https://www.gp.org/green_new_deal

27. European Commission. *A European Green Deal.* https://ec.europa.eu/info/strategy/priorities-2019-2024/european-green-deal_en

28. Clive Hamilton (1997). Foundations of ecological economics. In: Mark Diesendorf & Clive Hamilton (eds). *Human Ecology Human Economy: Ideas for an ecologically sustainable future.* Allen & Unwin, chapter 2.

29. Kerryn Higgs (2021). How the world embraced consumerism. https://www.bbc.com/future/article/20210120-how-the-world-became-consumerist

30. Herman Daly (1977). *Steady-State Economics: The economics of biophysical equilibrium and moral growth*. W. H. Freeman.

31. Philip Lawn (2006). An assessment of alternative measures of sustainable economic welfare. In: Philip Lawn (ed.) *Sustainable Development Indicators in Ecological* Economics. Edward Elgar Publishing, pp. 139–165; Eurostat (n.d.) https://ec.europa.eu/eurostat/statistics-explained/index.php?title=Quality_of_life_indicators_-_measuring_quality_of_life

32. Daly, *op. cit.*

33. There are at least three definitions of the Global South—see for example, Marlea Clarke https://onlineacademiccommunity.uvic.ca/globalsouth-politics/2018/08/08/global-south-what-does-it-mean-and-why-use-the-term/. In our book we use the term as defined in the main text and glossary, with the intentional overtone of subjugation. Alternative terms, for example, 'developing' and 'less developed' are also subjective as they suggest that 'developed' (i.e. rich) countries are a suitable model for 'development'.

34. Charles Phillips (2007). *The Illustrated History of the Aztec & Maya*. Lorenz Books.

35. Henry Louis Gates. *Africa's Great Civilisations*. DVD & Bluray. PBS. https://shop.pbs.org/WA7482.html

36. Timothy Brook (ed.) (2007–2009). *History of Imperial China*. Six volumes. Belknap Press

37. Frantz Fanon (2004). *The Wretched of the Earth*. New York: Grove Press; Aimé Césaire (2001) *Discourse on Colonialism*. New York: Monthly Review Press; Chinua Achebe (2017). *Things Fall Apart*. Penguin; Rodney Walter (2014). *How Europe Underdeveloped Africa*. Revised ed., African Tree Press.

38. Robert Waddell (1993). *Replanting the Banana Tree: A study in ecologically sustainable development*. University of Technology Sydney: APACE, chapter 1.

39. https://www.worldatlas.com/articles/what-is-decolonization.html

40. Kwame Nkrumah (1965). *Neo-Colonialism, the Last Stage of Imperialism*. https://www.abibitumi.com/wp-content/uploads/pp Migration/42994=1853-NeocolonialismThe-Last-Stage-of-Imperialism.pdf

41. Waddell *op. cit.*; Godfrey Uzoigwe (2019). Neocolonialism is dead: long live neocolonialism. *Journal of Global South Studies* 36(1):59–87; Naomi Klein (2008). *The Shock Doctrine: The rise of disaster capitalism.* Penguin.
42. Wolfgang Sachs (1992). *The Development Dictionary: A guide to knowledge as power.* London: Zed Books, p. 4.
43. Vandana Shiva (1989). *Staying Alive: Women, ecology and development.* London: Zed Books, p. 1.
44. Shiva, *op. cit.*
45. Charles McKelvey (2014). The erosion of neocolonialism. https://www.globallearning-cuba.com/blog-umlthe-view-from-the-southuml/the-erosion-of-neocolonialism
46. Wiedmann et al. (2020), *op. cit.*, Table 1.
47. Walter Miller (1959). *A Canticle for Leibowitz.* Lippincott & Co.
48. Oliver Balch (2013). Buen vivir: the social philosophy inspiring movements in South America. *The Guardian.* https://www.theguardian.com/sustainable-business/blog/buen-vivir-philosophy-south-america-eduardo-gudynas; Ashish Kothari et al. (2014). Buen vivir, degrowth and ecological swaraj. *Development* 57(3–4):362–375. https://doi.org/10.1057/dev.2015.24; Rapid Transition Alliance (2018). Buen vivir: the rights of nature in Bolivia and Ecuador. https://www.rapidtransition.org/stories/the-rights-of-nature-in-bolivia-and-ecuador/
49. Kothari et al. *op. cit.*

Reference

Weblinks accessed 26/10/2022.

4

Transitioning the Energy System

We are like tenant farmers chopping down the fence around our house for fuel when we should be using Nature's inexhaustible sources of energy—sun, wind and tide. I'd put my money on the sun and solar energy. (Attributed to Thomas A. Edison)

4.1 Introduction

Often the most difficult things to see are those surrounding us. One of those is energy. Although energy itself is invisible, it is manifest in many forms, some obvious, such as the chemical energy stored in the in a car's petrol tank or in the electrical energy illuminating our homes. Other forms are more subtle, hidden within everyday objects such as the energy required to make the building materials of our homes or in tasks such an online search or a surgical operation in a hospital.

Many people live in a global industrial economy that cannot function without access to reliable, affordable, abundant energy. Our civilisation has been largely built on our ability to exploit the energy stored in fossil fuels (FF). Coal was the driver of the Industrial Revolution, oil fuelled the motor car and natural gas became the principal source of heating and the basis of the chemical industry. But, as our consumption of 'goods' and services and our population increased, so did our use of FF and its adverse effects.

The combustion of FF produces heat, some of which is used in our homes, and some of which drives the wheels of industry and transport. But combustion emits pollutants of various types. The greenhouse gas (GHG) carbon dioxide (CO_2) remains in the global atmosphere for centuries, acting as an invisible blanket that allows sunshine to pass through while trapping heat that tries to escape. This has resulted in over-heating the Earth and changing its climate in life-threatening ways. Local air pollutants from FF combustion include oxides of nitrogen and sulphur, carbon monoxide, mercury, fluoride and particulates. These are toxic and/or carcinogenic and/or environmentally damaging.[1] The World Health Organization estimates that air pollution, outdoor and indoor combined, is responsible for about eight million deaths globally per year.[2] Most of these deaths result from the combustion of FF. The mining of FF is responsible for land degradation and the pollution of rivers, lakes and the water table. The prices of FF fluctuate substantially according to supply and demand, causing socio-economic stress to families and businesses. FF are concentrated sources of energy, used to produce electricity from huge power stations, and petrol and diesel from huge oil refineries. The owners and supporting industries of these facilities have gained substantial political power, creating dependence of individuals and communities upon large corporations and external governments. As we write, the European Union is working hard to reduce its dependence on imports of FF from Russia.

Therefore, in the interests of our health and wellbeing, and that of the whole biosphere, we must shift to energy sources and energy uses that have very low adverse environmental, health and social impacts. The Sustainable Civilisation, introduced in Chap. 3, is based on an ecologically sustainable energy system. On a global scale, a fully transitioned system comprises an energy supply that's sourced entirely from renewable energy (RE) together with substantial reductions in energy consumption. A reduction in high-income countries can more than offset the increased energy supply in low-income countries necessary for access to basics of human needs, such as clean water, food, housing and sanitation. The majority of RE is provided in the form of renewable electricity (RElec)[3] by solar photovoltaics (PV) and wind. These are the cheapest electricity generation technologies, even after storage has been added.[4] In several regions, hydroelectricity continues to play a major role, but there are few new large-scale developments due to resource and environmental constraints.

Demand for energy can be reduced by both improved technologies (Sect. 4.2) and behaviour change (Sect. 4.4). The former involves the efficient generation, conversion and use of energy. The latter can be encouraged by government policies to provide better infrastructure—that is, urban design, building codes and public transport—and policies to foster socio-economic changes, including constraints on the quantity and types of consumption.

The principal target for energy in the Sustainable Civilisation is zero anthropogenic (human-induced) GHG emissions and greatly reduced air and water pollution. Additional advantages are zero fuel prices (except where bioenergy is used), reduced use of water and less dependence upon large, politically powerful, energy corporations and potentially hostile energy exporting countries. However, it does entail increased use of materials, some of which are in limited supply (see Chap. 5).

This chapter has three aims. The first is to show that the transition to an energy efficient socio-economy powered entirely by RE is technically feasible and affordable, although it requires up-front investments in both money and energy. The technological transition to 100% RE (100RE) is discussed in Sect. 4.2 and the key policies needed to implement it are discussed in Sect. 4.3.

The second aim is to refute the myths disseminated by vested interests to undermine confidence in RE and EE—together they comprise ecologically sustainable energy. These myths are one of the main barriers to the transition to sustainable energy—they are refuted in Sect. 4.4.

The third aim is to show that, if energy consumption returns to pre-COVID growth rates, then it's unlikely that this transition can be completed in time to keep global heating well below 2 °C, the Paris Agreement's climate change target.[5] For readers coming from a purely technological perspective, this may come as a shock, but unfortunately it's true. For a *rapid* transition, we need both technological and, to reduce consumption, socio-economic changes, as discussed in Sect. 4.5.

Section 4.6 is the summary and conclusion. Readers may prefer to read Sect. 4.6 next, and then return to Sects. 4.2, 4.3, 4.4 and 4.5 that provide the supporting evidence and arguments.

4.2   The Technological Transition

Transition Strategy

The proposed strategy for transitioning the world from FF to RE follows from the following three propositions that are supported by almost all energy experts who are not connected to the FF or nuclear industries:

- Wind and solar PV can supply the vast majority of electricity demand at prices below that of new fossil-fuelled electricity and far below that of new nuclear power stations (see references cited in Sect. 4.1 and the subsection on nuclear energy below).
- Almost all residential, commercial and industrial heating that is currently supplied by burning fossil fuels (mainly gas) could be supplied economically by electric heat pumps, such as reverse cycle air conditioners, and direct electrical heating.
- Within several years, electric vehicles (EVs) are likely to be economically competitive with internal combustion engine (ICE) vehicles for general transportation purposes. The vehicle transition is being accelerated in several countries by means of emission standards for fleets of ICE vehicles, tax incentives and/or a rollout of fast-charge stations.

Therefore, the strategy for the global energy transition is as follows: to replace fossil-fuelled electricity with RElec supplied mainly by wind and solar PV; to build additional storage where required; to electrify all heating, apart from solar hot water and passive solar design of buildings; to electrify almost all road transportation; to reduce the cost of 'green' hydrogen from RE for long-distance aviation and shipping and non-energy industrial uses such as steel-making; and to greatly increase the energy efficiency of buildings, appliances and industrial processes.

The technologies for this strategy are all commercially available and most are already economically competitive. Only green hydrogen requires further development leading to cost reduction. Green hydrogen can be produced either by using RElec to split water by electrolysis or by using concentrated solar technology to produced high-temperature heat to split water thermally.[6] Because hydrogen is a difficult gas to manage and contain, it is likely that it would be used in the form of ammonia produced by combining hydrogen with nitrogen from the air. Even before green hydrogen or green ammonia becomes economic, the above strategy would eliminate emissions from electricity generation, heating, most transportation and almost all fugitive emissions,[7] amounting to about three-quarters of global GHG emissions.

The first dot point is supported by scores of computer simulation models for 100RElec and 100RE[8] in individual countries, regions (e.g. Europe) and the whole world. Two of the leading modelling research groups are based at Lappeenranta University in Finland (e.g. Dimitrii Bogdanov et al. 2021[9]) and at Stanford University in California (e.g. Mark Jacobson et al. 2015[10]). The scenario modelling is already being confirmed by practical experience in several countries and states (Table 4.1).

Before discussing the technologies and policies needed to facilitate the energy transition strategy, we must consider the situation at our starting point.

The Current Situation

In our homes, offices, shops, factories and industries, we use energy in three forms: as electricity, as heat supplied mostly by gaseous and liquid fuels, and as transport fuel. These forms of energy provide the energy services we demand: for example, warm homes in winter, hot showers, cold food, lighting and transportation.

Figure 4.1 shows the flows of global commercial energy purchased by households, offices, businesses and industries. (It doesn't count the free sunshine that makes the Earth habitable and grows our food.) Although the useful energy flow is from left to right, the driving force of that flow acts from right to left. Within the energy sector, the demand for energy services is the key driver. This in turn drives the demand for, and consumption of, energy at the point of use, the end-use energy or total final energy consumption (TFEC). Global trends in TFEC are discussed in

Table 4.1 Status of variable renewable electricity in selected countries and states, 2019

Country	Renewable electricity as % of total in 2019	Comment
Denmark	RElec 69% of consumption comprising VRElec 50% (mostly wind) + bioenergy from agricultural wastes 19%	100% RElec expected in 2028 Denmark is connected to Norway's hydro system
Scotland	RElec 56% of generation (VRElec 40%, hydro 10%) and 76% of consumption (97% in 2020)	VRElec mostly wind, supplemented by solar PV; big electricity exports to England
Germany	RElec 46% of public generation (wind 25%; PV 9%; biomass 9%; hydro 4%)	Target: At least 80% RElec by 2050
Ireland	RElec 42% of generation in 2020; wind 37%.	Negligible net trade in electricity
China	RElec 26% (wind 4.8%; solar 3%; hydro 17.8%)	2019 data: Thermal (mostly coal) is still 69% & nuclear 4.8%;
State		
Australian state: South Australia	RElec 50% of generation (mostly wind + rooftop PV); 69% in 2022	State gov't expects 100% net RElec by 2030. South Australia has only a weak transmission link to a neighbouring state; a new link is planned
German states: Schleswig-Holstein & Mecklenburg-Vorpommern	About 100% net generation each— Mostly wind	High-capacity transmission links with their neighbours balance variability of VRElec
US states: Iowa, Kansas & Oklahoma	42%, 41% & 35% generation from wind respectively	Texas with 17.5% has largest wind capacity in megawatts

Source: The author's compilation from many sources
Notes: RElec is renewable electricity. VRElec is variable renewable electricity (wind and solar PV). Only countries and states where hydro plays a minor role, if any, are listed

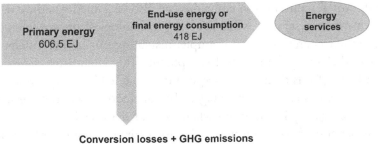

Conversion losses + GHG emissions
188.5 EJ = 31%

Fig. 4.1 Flow of global energy from primary sources to end-use to energy services, 2019. (Source: IEA data[11])

Box 4.1. TFEC in turn drives total primary energy supply (TPES), which determines most of the emissions. Clearly, a reduction in the demand for energy services—for example, heating your home to a lower temperature and wearing warm clothes; walking instead of driving to the local shops; taking shorter showers—can result in a large reduction in primary energy and hence emissions. These changes in behaviour are called *energy conservation* or *energy sufficiency*.

Box 4.1 Trends in Global Energy Consumption

Energy at the point of use is called *end-use energy* or *final energy* or *total final energy consumption* (TFEC). On a global scale, TFEC has been increasing approximately linearly since year 2000, apart from temporary small decreases during the Global Financial Crisis in 2009[12] and during the first year of the COVID-19 pandemic, 2020.[13] The good news is that the growth in energy consumption is not exponential. The bad news is the large size of the linear growth, 42.5% from 2000 to 2019, and the observation that energy growth was recovering in 2021[14] and could return to the pre-COVID trend. This will slow the transition to 100RE.

Global energy-related CO_2 emissions have increased at an even higher rate than TFEC, 44.6%, over the period 2000 to 2019. This rate is higher because most emissions originate from *primary energy supply*, that is, energy in natural forms—e.g. coal, oil, natural gas, wind, solar—before it's converted to useful final or end-use energy. The conversion process results in substantial energy losses, particularly in the form of waste heat in power stations, oil refineries and plants that compress and liquefy natural gas. To these must be added the so-called *fugitive emissions*, that is, losses of the GHG methane from the oil and gas fields, coalmines and gas pipelines.

Substantial reductions in emissions can also be gained when we retain the same energy services while applying technological improvements to reduce their energy uses, for example, insulating our home; purchasing energy efficient appliances and equipment. This is end-use *energy efficiency* (EE). (EE is defined as energy output divided by energy input and is usually expressed as a percentage.) The distinction between conservation and efficiency is important when we consider policies.

The benefits of acting on the driving forces are much greater for reducing the consumption of FF *electricity*. Burning a fuel to generate electricity is a very inefficient process, as shown in Fig. 4.2. Depending on the type of thermal power station and fuel combusted, the efficiency ranges from 25% to 50%. Expressing it another way, saving one unit of FF electricity by energy conservation and/or EE, or replacing it with one unit of RElec, saves two to four units of FF primary energy and their GHG emissions. Since the transition strategy discussed in the next section is based on the electrification of almost all energy use, the result of this transition will be a substantial reduction in primary FF energy and hence emissions.

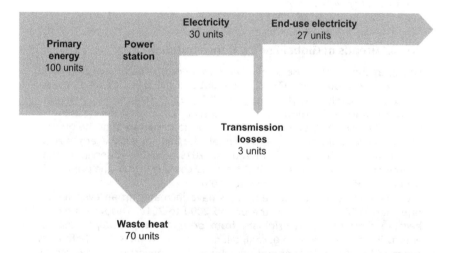

Fig. 4.2 Typical flow of energy in electricity generation by combustion. (Note: Conversion losses can range from 50% to 75% or more, depending on technology and fuel used. Average transmission and distribution losses are typically 10% in the Global North but can be much higher in the Global South. Source: the author)

Several countries and states are using their substantial hydroelectric resources to generate most of their electricity: Norway, Iceland (using geothermal power as well as hydro), New Zealand, Costa Rica, Uruguay, Paraguay, Nepal, Bhutan and Tasmania. However, hydroelectric resources are geographically limited, so most new grids transitioning to 100% RElec will have bulk electricity supplied by wind and/or solar PV. The transition from FF electricity to variable renewable electricity (VRElec, wind and solar PV) is already well advanced in several countries and states, as shown in Table 4.1. In addition, many cities, towns and industries have committed to 100% RElec.

Regions of the world with very large wind and/or solar energy resources include Australia, the USA, northern Europe, the Middle East, North Africa and northwest China. These regions can supply RElec by transmission line to regions lacking wind and solar resources.

In 2019, RElec comprised 26.2% of global electricity generation (hydro 15.9%, wind 5.3% and solar 2.5%)—see Fig. 4.3. The rapid growth of RElec has not slowed during the pandemic. Total global

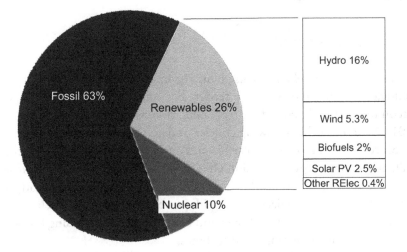

Fig. 4.3 Estimated renewable share of global electricity production, 2019. (Note: Numbers do not sum precisely, due to round-off error. The contributions of solid wastes are not shown. 'Other RElec' comprises geothermal, solar thermal and tidal electricity. Source: IEA data[17])

installed capacity grew by 256 gigawatts (GW) (almost 10%) from the end of 2019 to the end of 2020 where it reached 2839 GW and generated an estimated 29% of global electricity.[15] Since 2014, global annual investment in new RElec capacity has been greater than US$250 billion; in 2020, it was US$303.5 billion,[16] that is, three times the combined annual investment in fossil-fuelled and nuclear electricity capacity. In 2020, about 75% of net annual additions to global electricity generating capacity were renewables. Therefore, renewable electricity is decreasingly called 'alternative energy'. It is mainstream. This is all good news, at least for the electricity component of the energy transition.

But electricity is only 22% of global TFEC. The biggest challenge is to electrify transport and non-electrical heating rapidly, while increasing the rate of replacement of FF electricity with RElec. This will be a slow process while global energy consumption continues to increase and the growth of transport energy and of a large proportion of heating energy is still fossil fuelled (see Sect. 4.5).

As the farmer said to the hikers, "I wouldn't start from here." But we have no choice but to find a path from the state of the Earth in 2023 to Mount Sustainability.

Technologies for RE

There are many excellent books giving the basic information on RE technologies for the non-specialist reader. For scientists and engineers, we recommend the latest edition of Bent Sørensen's book published in 2017[18] and offer the following concise update to a rapidly evolving field.

Since 2017, the costs of solar PV and onshore wind have continued to fall, and their energy conversion efficiencies have continued to improve. In some regions of the world, these VRElec technologies are already so cheap that they can replace existing FF power stations before their retirement dates. The costs of offshore wind have also been decreasing for both fixed and the new floating wind turbines.

Big batteries have begun to play important roles in the grid. They can supply storage for periods of several hours to help smooth the variability of wind, solar PV and demand. Similarly, they can reduce the peak demand on transmission and distribution lines and so defer the need for upgrading or replacing the lines. When there are sudden, unpredictable variations in demand and supply, batteries can respond in a fraction of a second, much faster than any peak-load power station. They can help restart the system after a system-wide failure (blackout).[19]

As EVs take over from ICE vehicles, a huge amount of battery storage will become available. The challenge will be to implement an electricity pricing system and infrastructure to encourage motorists to charge their batteries while the Sun is shining and to discharge their batteries into the grid when electricity demand is high.

For storage of VRElec over periods of days or longer, batteries will still be very expensive for the foreseeable future. *Pumped hydro*, the principal form of storage of electrical energy in the grid for the past century, is cheaper for energy storage over days or longer. Since conventional pumped hydro on a river is limited geographically, it has been proposed that many thousands of sites around the world could be developed for small off-river pumped hydro plants.[20]

In cold climates, such as northern Europe, seasonal storage will be needed for 100% RElec systems to supply the higher electricity demand in winter. Hydroelectricity will be supplemented by using excess VRElec, that would otherwise be curtailed, to produce green hydrogen. These systems have low round-trip efficiencies—only about half the initial input energy can be recovered. Research and development are working to increase the efficiency and reduce the cost of electrolysis and also to further develop the thermal decomposition of water. More generally, the use of excess VRElec for other purposes than stored electricity—such as the production of green hydrogen for heating, steel-making, fertilisers and other chemicals—is known as *Power-to-X* or *P2X* and plays an important role in some 100RE scenarios.[21] Storage, in the context of maintaining the reliability of electricity supply, is discussed in Sect. 4.4.

Renewable Energy for the Global South

For the Global South, we first discuss two extreme renewable energy scenarios. One, a recent scenario study for Africa, is characterised by large-scale solar, supplemented by large-scale wind and hydro, linked by a network of high-voltage transmission lines criss-crossing the continent, together with green hydrogen. Substantial advances are assumed in battery storage and power-to-gas technologies. This is very much the Western development model; the cited study is co-authored by Western experts.[22] However, one would have to ask, "whom would such a system serve and who would pay for it?" The notion that large-scale centralised renewable energy will somehow lift the low-income majority of the Global South out of poverty is contested to say the least. The evidence suggests that most of the benefits will flow to the urban rich, the owners of industry, and big rural landowners.[23]

The other extreme is a less expensive, more accessible model to serve the low-income majority, who may never be connected to an electricity grid. It is based on the widespread use by households of a single solar PV panel together with a battery with sufficient capacity to power lights, a radio, a mobile phone and, for some farmers, solar water pumping. This is all that is needed for powering a village home or a small business. Household solar hot water is an optional extra for those who can afford it. Traditional forms of biomass energy are still needed for cooking, which requires a lot more energy. A variant of the single household system is a hub of solar panels and a bank of batteries to supply a whole village. When a household needs electricity, they remove a battery from the bank, use it at home and return it later for charging.[24] Both these off-grid models are spreading across low-income parts of Africa, south Asia and the Pacific islands.[25]

In practice, a combination of both models is needed to provide for the rapidly growing populations of cities and industrial developments as well as for rural subsistence populations, in the Global South. Meanwhile decarbonisation of the existing large-scale electricity grids in Africa and south Asia must be a priority.[26] However, as climate change progresses, transmission lines are becoming increasingly vulnerable to floods and wildfires, in rich countries as well as poor. Following the devastating Black Summer firestorms in Australia in 2019–20 (see Box 5.4), many rural people who lost electricity supply from the grid have been installing household solar.[27]

Technologies for Energy Efficiency

The two most effective GHG mitigation scenarios of the International Energy Agency (IEA) are its Sustainable Development Scenario and its Net Zero Emissions 2050 scenario.[28] Both scenarios have major contributions from efficient energy use.

For heating buildings, a major energy consumption, the shift to *heat pumps*, such as reverse cycle air conditioners, will give huge improvements in EE compared with heating with FF: one unit of electrical energy input can deliver 3–5 units of heat energy, depending on the device and the local weather. This doesn't violate the conservation of energy because the electricity input is used to 'pump' heat into the building from outside, not to heat the building directly. The main barrier to this emissions-reducing switch is the investment many households have made previously in gas heating—they may be reluctant to make this a stranded asset.

The transition from ICE vehicles, with typical EEs of 15–30%, to EVs, with EEs of 75–85%, also provides huge gains in EE. In this case we have defined EE as energy applied to the wheels divided by energy used from the fuel tank or battery. If the electricity supplied to an EV is fossil fuelled, its overall efficiency advantage is reduced substantially, taking account of the low efficiencies of thermal power stations. To maintain an EV's huge efficiency advantage and zero GHG emissions during operation, it must be 'fuelled' with renewable electricity.

An existing example of highly efficient energy generation of renewable energy is Denmark's use of combined heat and power from the combustion of agricultural residues and municipal wastes.[29] The waste heat from electricity generation is used for district heating. The efficiency of using the fuel is over 85%.[30] More generally, industrial 'parks' where energy cascades down from factories that require high-temperature heat (e.g. steel-making and aluminium) to those that require low-temperature heat (e.g. food processing) is a highly efficient way of using a fuel. The Kalundborg Eco-Industrial Park in Denmark is a classic example where both energy, water and solid material by-products are used cooperatively between different industrial processes.[31]

Does Nuclear Energy Have a Future?

If we are to believe its proponents, nuclear energy is undergoing a renaissance. However, based on actual data, this claim does not stand up to examination. Nuclear energy's share of global electricity generation has declined from its peak of 17.5% in 1996 to 10.1% in 2020. Its net global electricity generation in 2020 was 2553 terawatt-hours (TWh), which is just under its peak of 2660 TWh in 2006. Annual investment decisions for the construction of new nuclear reactors in 2020 were US$18.4 billion for 5 gigawatts (GW) of generating capacity, compared with the total of US$303.5 billion for wind (new capacity 73 GW) and solar (new capacity 132 GW).[32]

According to multinational investment advisor Lazard, electricity from new nuclear power reactors in the USA would cost about four times that from wind or solar PV farms.[33] So far, the modest global annual investment in nuclear energy is in types of reactors that offer small improvements in existing designs, but are much more expensive.

Although the renaissance has not arrived, the nuclear industry and its supporters claim that a new generation of reactors, allegedly safer and cheaper, will soon become commercially available. Most publicity is given to so-called 'small modular reactors' (SMRs), whose theoretical advantages are that they could be mass-produced in factories and hence would generate cheaper electricity than existing large reactors. While small reactors are used in nuclear-powered military submarines, they are very expensive, unsuitable for mass-production, and cannot be described accurately as modular. Despite much hype, there are no orders for multiple SMRs. Therefore, if they ever became commercially available with the assistance of government subsidies, their electricity would likely be even more expensive than from big, conventional nuclear reactors. SMRs have been described as 'paper reactors fuelled on ink'. Although much research, development and demonstration funding has been devoted to other types of reactor, such as the fast breeder and the thorium reactor, none is commercially available.

The long-standing arguments against nuclear fission power are still relevant today: nuclear power is too expensive; it is likely to have occasional devastating accidents like Chernobyl and Fukushima; it contributes to the proliferation of nuclear weapons; it produces dangerous wastes that must be managed for hundreds of thousands of years; it is too slow to build to help mitigate the climate crisis; and it is insufficiently flexible in operation to be a good partner for wind and solar power.[34] In January 2022, the former heads of nuclear regulatory bodies across Europe and the USA released a statement voicing their opposition to nuclear energy as a climate solution on these grounds.[35] Furthermore, global reserves of high-grade uranium ore are limited. When low-grade ore has to be mined and milled, energy inputs will become substantial.[36] The status of nuclear power is summarised in Box 4.2.

Box 4.2 Nuclear Power in a Nutshell

Too expensive; too slow to build; too dangerous; too inflexible in operation; too energy intensive.

Nuclear fusion has not been achieved in a controlled manner on Earth for more than several seconds under conditions when energy output is greater than energy input. If the current expensive international project, ITER[37] eventually succeeds in overcoming this barrier, the technology could not become commercially available before the mid-2040s, too late for the climate.

The policy implications are that countries without nuclear power would be wasting their resources and increasing the risk of nuclear weapons proliferation by investing in nuclear power. Countries with ageing nuclear power stations should replace them with RElec or, if local resources are insufficient, transmission links enabling them to import RElec.

4.3 Policies for Sustainable Energy

The market is a slow and often an ineffective means of building new essential infrastructure. To assist the rapid transition to 100% RElec in which most electricity is generated by VRElec, the following key government policies are needed.[38]

Transmission Lines, Renewable Energy Zones and Storage

Wind and solar farms can be planned and built in three years. Because they cannot all be located at the sites of retired FF power stations or close to existing transmission lines, new transmission lines may be needed. For example, in Europe, a new major link is needed between the sunny south and the windy north; in the USA, windy Texas must be joined to neighbouring states; in Australia, the state of South Australia, which is likely to reach 100RElec before 2030, needs a link to the coal dependent state of New South Wales. This link is belatedly under construction. A major new transmission line can take two to three times as long to build as a solar or wind farm, partly because planning permission and easements must be obtained for the line, and many privately owned properties may be involved. (However, upgrading a line on an existing easement can be rapid.)

If a site is suitable for multiple wind and solar farms and storage, it makes economic sense and saves time if governments plan a *Renewable Energy Zone* (REZ) at that site, with a single transmission line joining it to the main grid.[39]

Government policies are required to plan transmission lines, REZs and storage in consultation with local communities, RE developers and the electricity industry, and, at bare minimum, fund grants for feasibility studies and local site preparation.

Reform of Electricity Markets

Electricity markets were designed at a time when large-scale electricity systems comprised a relatively small number of large power stations feeding electricity in one direction to consumers in households, businesses and industries. Now the situation is changing dramatically to grid systems, each with millions of RElec power stations ranging in generating capacity from kilowatts to gigawatts, and two-way flows of electricity. Consumers are becoming *prosumers*, that is, households and organisations that sell self-generated electricity back to the grid as well as purchasing electricity from the grid. In addition, unlike FF power stations, wind and solar farms have very low operating costs, a situation that can result in negative wholesale prices during sunny days.

Therefore, electricity markets need new rules for the new operating environment. Inter alia, the rules should ensure that demand response, batteries and other forms of storage are properly rewarded for their contributions and that VRElec generators receive a sufficient average wholesale price to pay off their capital costs. The climate crisis demands that markets have a climate objective to supplement the existing economic and reliability objectives.

Removing Subsidies to FF

Estimates of the actual amounts depend on the definition of 'subsidy' used and the scope of the data collected. The International Institute for Sustainable Development estimates that subsidies to FF by the G20 governments alone amount to US$584 billion per year (2017–2019 average) via direct budgetary transfers and tax expenditure, price support, public finance, and investment by state-owned enterprises for the production and consumption of FF at home and abroad.[40] The International Energy Agency's estimate for direct global subsidies in 2021 is US$440 billion, with the top five countries being Iran, China, India, Saudi Arabia and Russia, in that order.[41] These subsidies distort markets, encourage wasteful consumption, disproportionately benefit wealthier households, increase the environmental and health costs of FF use, and discourage the adoption of renewable energy (RE). These government subsidies are a result of corporate capture of nation-states, discussed in Chap. 6.

Pricing Carbon

Burning FF imposes environmental and health costs onto the community. The failure to include these costs in FF prices is an indirect subsidy. Therefore, governments should apply a carbon price to FF that reflects, at least roughly, their external costs. We prefer a carbon tax, because it's easy for a poorly designed emissions trading scheme (ETS) to fail. This was the situation with the first two stages of the European ETS.[42]

Rapid Phase-out of Coal with Social Justice for Workers

The necessary rapid growth of RElec must be matched with the rapid retirement of coal-fired power stations. Older coal stations are gradually being retired in regions where wind and/or solar, with very low operating costs, are displacing the continuous operation of coal plants. The German government has been encouraging the planned closure of black coal power stations by means of a series of tenders between 2020 and 2021.[43] Germany's brown coal stations are being closed under a separate scheme with fixed compensation. To speed up the retirement of coal power, the following policy could be adopted: plants bid the payment they require for closure and the regulator chooses the most cost-effective bid; the plants remaining in operation then make financial transfers to the plant that exits, in line with their emissions, under government regulation. This policy would cost the government nothing.[44] Fossil fuels could be phased out even more rapidly by placing global and national caps on their extraction, with tighter caps for the Global North. Needless to say, this would be difficult politically.

The social justice aspect of this transition is government policy to assist workers in coal mining regions to retrain, relocate or retire on an adequate pension. Governments can also give incentives for new, cleaner industries to establish in these regions. Spain is closing all its coal-fired power stations and has negotiated with trade unions a Just Transition deal guaranteeing both early retirement and massive investment to create alternative jobs.[45]

Encouraging Energy Efficiency

EE is another issue where the market fails, especially with rented accommodation. Government policy intervention is needed to set energy performance standards for buildings, appliances and other energy-using equipment. Energy labelling should be mandatory. Energy audits should be required for residential and commercial buildings and the results should be displayed on sale and rental contracts. Feebates and rebates can be introduced for appliance sales: a surcharge on inefficient appliances (the feebate) pays for rebates on efficient appliances. Government incentives are needed to create eco-industrial 'parks', where the energy and material 'wastes' from some industries become resources for other industries.

Before identifying the major challenges to the energy transition, we examine critically the principal myths that have been circulated by vested interests and their supporters to delay action.

4.4   Busting Myths about Sustainable Energy

The Base-Load Myth

First, we explain the context of this myth. Electricity supply and demand must be balanced continuously in an electricity grid. If demand exceeds supply, even slightly, the frequency of oscillation of the alternating current (AC) falls. Conversely, if supply exceeds demand, the frequency increases. In either case, electrical equipment that's designed to receive AC at the specified frequency[46] can be damaged if the frequency deviation exceeds quite small bounds. If the supply and demand become substantially out of balance, a blackout occurs.

Demand varies throughout the day, typically with a large peak in the early evening. Traditionally, grid operators varied supply to follow demand. They managed this by having two principal types of power stations, base-load and peak-load. Most base-load power stations are fuelled on coal (or uranium in several countries) because gas is more expensive for continuous operation and hydroelectricity is limited to certain regions. Base-load power stations perform best, and are most economical, when they operate continuously 24/7 at their rated power output. But, except for base-load hydro, they are inflexible in operation, in that they cannot vary their power output rapidly to meet changes in demand. This task is performed by peak-load power stations, either hydro or open-cycle gas turbines (essentially jet engines), which are only operated for short periods to supply peaks in demand or temporarily to help fill a gap in supply when a base-load power station unexpectedly breaks down.

Nowadays large-scale electricity systems are transitioning to 100% RElec, with the bulk of electricity supplied by wind and solar PV. Vested interests and their supporters claim that base-load power stations are still necessary for reliable supply and that RElec cannot provide them—this is *the base-load myth*, which is still being repeated by some energy 'experts', politicians and journalists.

Although efforts are being made to make base-load power stations more flexible to 'balance' the variability of wind and solar, they cannot compete in response speed and flexibility of operation with peak-load power stations and other forms of energy storage. Storage for up to a few hours, the duration of peaks in demand, is increasingly being provided by big batteries. For periods of days, pumped hydro, thermal storage, compressed air and open-cycle gas turbines (which can be operated on renewable fuels) are suitable. Geographic dispersion of wind and solar farms and interconnection to neighbouring grids smooths the variability of RElec supply.

Furthermore, information technology is enabling demand to be modified rapidly when required. Businesses are beginning to organise large numbers of homes with solar power and batteries into *virtual power stations*.[47] These can help supply peaks in demand on the grid and withdraw electricity from the grid and store it when demand is low. In future, some electricity retailers may price electricity supplied to their household and small business customers in real time, according to supply and demand on the grid, thus giving customers incentives to undertake non-urgent tasks such as heating water and washing and ironing clothes during times of low demand on the grid.

Since wind and solar farms generally have smaller generating capacities than large thermal power stations, they enable a better match between new supply and changing demand, making generation planning and investment easier.

Contrary to the base-load myth, a well-designed large-scale electricity system, dominated by variable RElec, can be both reliable and affordable without base-load power stations. The base-load myth has been refuted by both simulation modelling[48] and practical experience (see Sect. 4.2) in electricity grids where VRElec supplies most annual electricity generation. The quantity of storage needed to maintain reliability depends on the penetration of VRElec into the grid and its local climate. Given high penetrations into the grid of VRElec, quite small amounts of storage are needed for Australia,[49] where typical storage periods are hours to a week. However, Europe is likely to also need seasonal storage, mainly in the form of Power-to-X (see Sect. 4.2).[50]

Frequency and Voltage Myth

This is a specialised technical topic that requires undergraduate knowledge of electrical engineering for full understanding. A short, simplified outline of how frequency and voltage are managed is offered in Box 4.3. A more detailed account is given by AEMO.[51]

Box 4.3 Managing Frequency and Voltage in a Grid with High Penetration of VRelec

A power system must operate in a stable manner through a wide variety of operating conditions. It must keep the voltage and frequency of AC at their set points and return to them quickly under regular minor changes, such as varying supply and demand, and also after severe disturbances such as an equipment failure or a power system fault due to lightning strike. In a traditional power system, the energy stored in the large rotating machines (turbines and generators) stabilises the power output when any mismatch occurs between generation and demand. The rotational speed of the machines sets the frequency and their inertia resists rapid changes.

Opponents of RElec sometimes claim that a power system cannot operate in a secure and stable manner without a high proportion of *synchronous generators*, that is, those that produce the same AC frequency as the grid, such as coal, gas or hydro machines.

This myth is refuted by noting that in the South Australian power system, with more than 60% of its annual electrical energy from solar and wind, *synchronous condensers*[52] ('syncons'),[53] with flywheels for added inertia, have been installed to deliver the services no longer provided by retired synchronous generators. They run at synchronous speed with high rotational kinetic energy and inertia without consuming any fuel.

Solar, wind, batteries and fuel cells are connected to the grid through an inverter, which converts direct current (DC) to AC at any specified voltage. The present generation of inverters, called *grid-following inverters*, supplies AC that's tied to the grid frequency. Without modification to their hardware and software, they cannot provide synthetic inertia and hence stabilise the grid at the required frequency. However, as the remaining fossil- fuelled synchronous generators retire, *grid forming inverters* that can simulate a rotating generator will be added to battery and solar systems to enable them to set the power system frequency. Smart converters will be added to wind generators to utilise their rotational kinetic energy and inertia to support a stable power system. With further development and mass-production, grid forming inverters and converters will replace syncons.[54]

Fossil Fuel Back-Up Myth

This myth claims that, even if we transition to 100% RElec, FF would still be needed for mining the raw materials and manufacturing the RE technologies—in other words, that the life-cycle inputs would still need FF. This argument is losing credibility as the mining and minerals processing industries begin to install their own solar and wind farms to provide most of their energy use[55] as RElec saves them money. Low-carbon steel making[56] and aluminium smelting[57] are under development and demonstration. Thus, the life-cycles of RE technologies are slowly but surely becoming independent of FF. Already several hundred large corporations have committed to 100% RElec by specified dates,[58] with some aspiring to 100% RE.

Net Energy Myth

In its most extreme form, this myth claims that RE requires more energy invested than it generates over its lifetime—in other words, it doesn't generate any *net energy*. A more common version of the myth is that RElec has low net energy or low *energy return on energy invested* (EROEI or EROI)[59] and so will compete excessively with investment needed in other sectors of the economy.

This is a cunning myth, because it was true several decades ago, when solar PV systems were made individually to power the early artificial satellites. But nowadays, PV modules are mass-produced and have much lower energy inputs, much higher energy conversion efficiencies and longer lifetimes. They generate the life-cycle primary energy required to construct themselves in 1–2 years, depending on their type and location, and their lifetime is at least 25 years.[60] A large wind turbine with similar lifetime generates the life-cycle primary energy required to construct itself in 6–12 months.[61]

Incidentally, FF electricity, as opposed to FF at the mine mouth, typically has net energy that's no better than that of solar PV.[62] That's mainly the result of the low efficiency of energy conversion when a fuel is combusted to generate electricity (see Fig. 4.2).

Some types of storage that are part of net energy generators (e.g. hydro-electricity with single dam; concentrated solar thermal with thermal storage) can increase EROI of the whole electricity system, while other forms that are not net energy generators (e.g. pumped hydro; battery) will decrease it.[63]

A separate but related net energy issue arises if the transition to large-scale RElec system occurs so rapidly that one batch of solar and wind farms is installed before the previous batch has generated the energy to build itself.[64] Then, temporarily, while the rapid transition is occurring, more energy must be invested in the whole electricity system than the energy it generates. However, after the transition is complete or slows down, energy generation will overtake and surpass energy investment. It must be emphasised that this situation is not the fault of RElec, because it could be worse if nuclear power, with its very long construction time, were used instead of RElec to substitute for FF electricity. The responsibility for the issue belongs to industries and governments that have delayed the transition for so long that now a very rapid transition is needed to try to avoid climate tipping points.

Land-Use Myth

This myth is that wind and solar farms need vast areas of land, thus competing with food production and threatening biodiversity. The myth is sometimes expressed in terms of RE having 'low energy density' or 'low power density'.

Most wind and solar farms are built either on agricultural land or on marginal land. Although wind farms span large areas, the turbines and access roads occupy just a tiny fraction of the land spanned, typically 1–2%.[65] Wind farms are compatible with essentially all forms of agriculture, giving farmers an additional source of income as rent. Some wind farms occupy less land than an equivalent coal-fired power station with open-cut coalmine. Offshore wind farms, which occupy no land, are becoming less expensive. Rooftop solar requires no additional land. Solar farms are compatible with some types of farming. Some are being built sufficiently high above the ground to provide shelter for sheep; others are even higher to allow crops to be grown and harvested. Partial shading shields crops from excessive sunlight, drought, storms and hail, and reduces irrigation needs. The sharing of land between agriculture and solar PV is known as *agrivoltaics*.[66]

Numerous studies have shown that there is ample land for 100% RElec globally, without competing with food production or desecrating national parks.[67] Regions deficient in RE resources, within and between countries, can and do import RElec by transmission line from countries with excess.

Misleading Measures of Energy Efficiency

Let's recall that EE was defined in Sect. 4.2 as having the same energy service with less energy consumption. A worrying trend is that some governments and economists have redefined EE to be either *energy intensity*, which is energy consumption divided by GDP, or *energy productivity*, which is GDP divided by energy consumption. The myth is that EE can be measured by these terms. But they are not genuine measures of EE, because energy intensity can decline and energy productivity can increase while energy consumption increases, so long as GDP increases more rapidly than energy consumption. Energy saving targets should be expressed in terms of reductions in absolute energy consumption.[68]

Rebound Myths

Rebound occurs in the energy sector when an investment is made in an EE or RE technology that saves the investor money. Then the investor spends the saved money on goods and services that use energy. For example, if I invest in an electric car, which has lower fuel costs than a petrol car, I may take longer trips than previously.

There are at least two myths associated with this form of rebound. The first is that the rebound is so large that it cancels the original savings in energy and CO_2 emissions. This myth has been used by some people to claim that energy efficiency and conservation are a waste of time and money. However, that's unlikely, because not all the money I save with EE and RE goes into energy use.

The second myth is that nothing can be done to reduce rebound. To the contrary, rebound can be reduced by policies such as implementing a carbon price; technical standards and labelling standards to foster products and services with low life-cycle energy consumption; identity signalling; feebates and rebates; and a government spending portfolio that avoids encouraging more consumption.[69] The latter is one type of policy that's consistent with a steady-state economy discussed in Chap. 7.

Materials Availability: An Exaggeration, but Not a Myth

One concern articulated by critics about the rapid growth of RElec must be taken seriously and acted upon: the risk that the demand for specific materials for RE and battery storage could exceed known reserves.[70] This is not a myth, but the risk has been exaggerated, because it has solutions: design for disassembly, recycling, substitution and improved efficiency of manufacturing (and, of course, by global reductions in energy consumption). Government policies are needed to encourage urgent action. This issue is not confined to the energy sector but is part of the much larger problem of the sustainability of material supply and use, discussed in Chap. 5.

Renewable Energy Deniers

People who cling to one or more of the above-refuted myths and ignore or reject the large body of evidence for the technical feasibility and affordability of 100RE are sometimes described as *renewable energy deniers*. They use similar tactics to climate science deniers and fall into (at least) three categories. First are the climate science deniers, who ignore all the environmental, health and economic benefits of transitioning to 100RE that are additional to stabilising human-induced climate change (see Chap. 1). Second is the nuclear lobby, which accepts climate science, but uses the base-load myth to argue that VRElec should have a limited role. Third is a group that accepts climate science and rejects 100RE (and, in some cases, rejects all modern RE and EVs) and believes in extreme degrowth to self-sufficient local communities using technologies from before the Industrial Revolution—see the debate in the peer-reviewed, open-access journal *Energies*.[71] We accept the need for the global reduction in the use of energy, materials and land, that is, degrowth in physical terms, discussed in the next section and in Sect. 7.3, but reject the poorly argued claims that human civilisation has to return to pre-industrial society.

4.5   Growth in Consumption: Chasing a Retreating Target

The previous section critiqued the principal myths, exaggerations and misleading definitions used by vested interests and their supporters to delay the transition to 100RE and greater EE. These alleged barriers are within the energy sector and have feasible solutions, mostly technical, within the sector.

The present section introduces a genuine major barrier to the *rapid* achievement of 100RE: the ongoing growth in global total final energy consumption (TFEC), which is in turn driven by growth in the global economy and population. Hence, this barrier is partly outside the energy sector. The problem can be stated simply by looking at global energy data. In year 2000, the percentage of total final energy consumption from FF was 80%; in 2019 it was still 80%.[72] How is this possible during a period when RElec has been growing rapidly?

Global energy consumption grew steadily during that period and most of the growth was fossil fuelled. If recent (2000–2019) rates of growth continue during the economic recovery from the pandemic, then it will be almost impossible for RE to overtake and replace all FF energy consumption by 2050. The problem is similar to that of a runner trying to break the record for her distance, while officials are moving the finishing tape away from her. She will reach the tape, eventually, but is unlikely to break the record.

It will be very difficult to rapidly replace all ICE vehicles by EVs and all combustion heating by electric heating while energy consumption for transportation and heating is growing. To reach 100RE by 2050, we must make these substitutions so rapidly that there is such high demand for RElec that it can grow exponentially. Although we can build wind and solar farms very quickly because the modules are mass produced in factories, RElec cannot grow more rapidly than the demand for electricity. A simple calculation shows that, in order for RElec to substitute for all FF energy—in electricity generation, transportation and heating—we need total global TFEC to decline substantially between now and 2050.[73] Such a decline is feasible, because of the gains in energy conversion efficiency between primary and end-use energy that will be achieved by transitioning to RE and EE. As discussed in Sect. 4.2, EVs and heating by electric heat pump are much more efficient than ICE vehicles and combustion heating, respectively. So, energy consumption in transportation and heating will be reduced automatically *provided* this reduction is not outweighed by increased demand for energy resulting from increasing economic growth. The challenges of achieving the rapid electrification of transport and combustion heating remain. A precautionary approach would aim to reduce global energy consumption by 50–75% by 2050.[74]

Next, we could ask, in which parts of the world is energy consumption growing and why is it growing there? IEA data show that TFEC has been declining since 2000 in the European Union as a whole and has been increasing only slightly in the Organisation for Economic Cooperation and Development (OECD) countries as a whole and in total North America. Does this mean that the rich countries are leading the way in reducing or at least stabilising energy consumption? Unfortunately, no! Although the really big increases are occurring in the rapidly growing economies such as China, India, Mexico, Brazil and Indonesia, this is partly because they are manufacturing 'goods' for the rich countries. Therefore, ending growth in global energy consumption must entail *reducing* the energy consumption of the rich countries as well as shifting the latter's imports to goods with low embodied carbon emissions. This will require increased economic cooperation between the rich and rapidly developing countries, and indeed the poor countries.

4.6   Summary and Conclusion

The key technical tasks in transitioning to an ecologically sustainable energy system are to convert all electricity generation to renewables, to electrify transportation and heating, and to greatly increase the efficiency of energy conversion and use.

The evidence that the Sustainable Civilisation could be powered entirely by renewable energy (RE) is two-fold. First, scores of scenarios, backed up by simulation modelling with hourly (or shorter) time-steps find that 100% renewable electricity (100RElec) is technically feasible and affordable for individual countries, regions and the whole world (see Sect. 4.2). Countries with insufficient RE resources can and do import RElec by transmission line. Wind and solar PV, which would supply most RE in the form of RElec, are now much cheaper than electricity from fossil fuels and nuclear energy. They are still cheaper after short-term storage has been added to electricity grids to balance their variability.

The only expensive technologies are batteries, which are becoming cheaper as markets grow, and 'green' hydrogen. The latter is needed for long-distance air and sea transport and road transport in remote areas, steel-making and several other non-energy industrial processes. Its cost will decline with continuing research and development, mass-production of electrolysers and possibly with advances in the thermal decomposition of water (an alternative way to make hydrogen). The key policies needed to drive the technological transition are well-known and easy to implement (see Sect. 4.3), given the political will. The technical myths disseminated by RE deniers do not stand up to examination (see Sect. 4.4).

The second body of evidence that the Sustainable Civilisation could be powered entirely by RE comes from practical experience that is confirming the results of simulation modelling. Several large-scale electricity supply systems, with half or more of their annual electricity generation supplied by wind and/or solar, are already operating reliably for periods of hours to days on 100RElec. The credibility of claims by RE deniers, that RE technologies will always need FF to construct them, is refuted by the ongoing transition of the mining, minerals processing and manufacturing industries to low-cost RElec for their electricity requirements. It's just a matter of time before the mining industry replaces diesel-fuelled earth-moving equipment with electric.

The real problem faced by the RE transition is neither technical feasibility nor affordability but timing. If global energy demand returns to pre-COVID growth rates (and this has been the trend in 2021–2022), then the growth in RE will be chasing a target that's running away from us, and it will become almost impossible for RE to replace all FF use by 2050. Then the probability that Earth's climate will cross a tipping point will increase greatly. To reduce this probability, total global final energy consumption must cease growing and moreover must decline to a greater degree than can be achieved by energy efficiency alone. Since low-income countries need more energy to develop, the principal responsibility for reducing energy consumption falls justly on the rich countries, which are responsible for the vast majority of excess CO_2 in the atmosphere.

Because energy consumption is strongly linked to economic activity, the rich countries must implement physical degrowth to a steady-state economy with reduced throughput, while greening their technologies and industries, and placing much more emphasis on improving the quality of life for all people. This entails a major change to the economic system, as discussed in Chap. 7. A related strategy that is needed to speed up the transition is to weaken the political power of the FF industry (see Chap. 6). Thus, energy transition needs strategies and policies that transcend technical fixes within the energy sector. The next chapter discusses the sustainability of other key natural resources.

Notes

1. Erica Burt et al. (2013). *Scientific Evidence of Health Effects from Coal Use in Energy Generation.* https://noharm-global.org/sites/default/files/documents-files/828/Health_Effects_Coal_Use_Energy_Generation. pdf; Paul R. Epstein et al. (2011). Full cost accounting for the life cycle of coal. *Ann. N. Y. Acad. Sci.* 1219:73–98. Doi: 10.1111/j.1749-6632. 2010.05890.x; K.R. Smith et al. (2013). Energy and human health. *Ann. Rev. Pub. Health* 34:159–188; NRDC (2018). *Fossil Fuels: The dirty facts.* https://www.nrdc.org/stories/fossil-fuels-dirty-facts

2. World Health Organisation. *Air Pollution*. https://www.who.int/health-topics/air-pollution#tab=tab_1

3. This chapter distinguishes between RE, which is renewable energy (all energy, not limited to electricity), and RElec, which is renewable electricity.

4. IRENA (2021). *Renewable Power Generation Costs in 2020*; International Renewable Energy Agency: Abu Dhabi, United Arab Emirates, https://www.irena.org/publications/2021/Jun/Renewable-Power-Costs-in-2020; Lazard (2021). *Lazard's Levelized Cost of Energy Analysis—Version 15.0*. https://www.lazard.com/media/451905/lazards-levelized-cost-of-energy-version-150-vf.pdf; Paul Graham et al. (2021). *GenCost 2020–21*. CSIRO: Canberra, Australia, https://doi.org/10.25919/8rpx-0666

5. United Nations Climate Change. *The Paris Agreement*. https://unfccc.int/process-and-meetings/the-paris-agreement/the-paris-agreement

6. Giles Parkinson (2022). *RenewEconomy*. https://reneweconomy.com.au/fossil-fuel-giant-woodside-turns-to-concentrated-solar-tech-in-new-hydrogen-play/

7. Fugitive emissions are GHG emissions from oil and gas fields, coal mines and leaks in gas pipelines.

8. These are our abbreviations for 100% RE and 100% RElec, respectively.

9. Dimitrii Bogdanov et al. (2021). Low-cost renewable electricity as the key driver of the global energy transition towards sustainability. *Energy* 227, 120,467. https://doi.org/10.1038/s41467-019-08855-1

10. Mark Z. Jacobson, Mark A. Delucchi MA et al. (2015) Low-cost solution to the grid reliability problem with 100% penetration of intermittent wind, water, and solar for all purposes. PNAS 112:15060–15,065.

11. IEA (2021a). *Key World Energy Statistics 2021*. https://www.iea.org/reports/key-world-energy-statistics-2021

12. IEA (2021b). *Key World Energy Statistics 2021. Final Consumption*. https://prod.iea.org/reports/key-world-energy-statistics-2021/final-consumption

13. IEA (2021c). *Global Energy Review 2021*. https://www.iea.org/reports/global-energy-review-2021

14. IEA (2021c). *op. cit.*

15. REN21 (2021). *Renewables 2021: Global Status Report*. https://www.unep.org/resources/report/renewables-2021-global-status-report

16. REN21 (2021), *op. cit.*

17. IEA (2021d). *Electricity Generation by Source –World*. https://www.iea. org/data-and-statistics/data-product/electricity-information

18. Bent Sørensen (2017). *Renewable Energy; Physics, engineering, environmental impacts, economics and planning*. 5th edition. Elsevier.

19. USAID & NREL (2019). *Grid-Scale Battery Storage: Frequently asked questions*. https://www.nrel.gov/docs/fy19osti/74426.pdf

20. Andrew Blakers, Bin Lu & Matthew Stocks (2017). 100% renewable electricity in Australia. *Energy* 133:471–482.

21. Bogdanov et al. (2021) *op. cit.*

22. Ayobami Oyewo et al. (2022). Contextualizing the scope, scale, and speed of energy pathways toward sustainable development in Africa, *iScience* (2022). https://doi.org/10.1016/j.isci.2022.104965

23. Fiona Burlig & Louis Preonas (2017). How large are the economic benefits of rural electrification? *ARE Update* 20(3):4–6. University of California Giannini Foundation of Agricultural Economics.https:// giannini.ucop.edu/filer/file/1487699545/17899/; Shahidure Khandker et al. (2014). Who benefits most from rural electrification? Evidence in India. *The Energy Journal* 35(2):75–96; Kenneth Lee et al. (2020). Experimental evidence on the economics of rural electrification. *Journal of Political Economy* 128(4):1523–1565; Peter Fairley (2016). Electrification causes economic growth, right? Maybe not. *IEEE Spectrum*, https://spectrum.ieee.org/does-electrification-cause-economic-growth

24. Planet Experts, http://www.planetexperts.com/intasave-is-empowering-african-communities-for-a-song/

25. Grameen Shakti, https://gshakti.org/; Tony Weir (2018). Renewable energy in the Pacific islands: its role and status. *Renewable & Sustainable Energy Reviews* 94:762–771. https://doi.org/10.1016/j.rser.2018.05.069; Anil Cabraal et al. (2018). *Living in the Light: The Bangladesh solar home systems story*. The World Bank, https://openknowledge.worldbank.org/ bitstream/handle/10986/35311/Living-in-the-Light-The-Bangladesh-Solar-Home-Systems-Story.pdf?sequence=1&isAllowed=y; A.K.E. Haque et al. (eds) (2022). *Climate Change and Community Resilience*. Springer. https://doi.org/10.1007/978-981-16-0680-9_14

26. Oyewo et al. *op. cit.*

27. Adam Morton (2020). *The Guardian*. https://www.theguardian.com/ environment/2020/feb/19/mike-and-annie-cannon-brookes-pledge-12m-to-supply-solar-systems-for-disaster-relief#:~:text=2%20 years%20old-,Mike%20and%20Annie%20Cannon%2DBrookes%20

pledge%20%2412 m%20to,solar%20systems%20for%20disaster%20
relief&text=Software%20billionaire%20Mike%20Cannon%2D
Brookes,grid%20by%20bushfire%20or%20flood

28. IEA (2020). World Energy Outlook 2020. https://www.iea.org/reports/world-energy-outlook-2020/overview-and-key-findings

29. There is some controversy about the environmental impacts of burning solid waste.

30. Danish Energy Agency, https://ens.dk/en/our-responsibilities/global-cooperation/experiences-district-heating

31. Engie Impact, https://www.engieimpact.com/insights/eco-industrial-park-case-study-kalundborg; InfoGalactic, https://infogalactic.com/info/Kalundborg_Eco-industrial_Park

32. Mycle Schneider & Anthony Froggatt (2021). *The World Nuclear Industry Status Report 2021.* Mycle Schneider Consulting, https://www.world-nuclearreport.org/-World-Nuclear-Industry-Status-Report-2021-.html

33. Lazard (2021). *op. cit.*

34. Benjamin Sovacool (2011) *Contesting the Future of Nuclear Power: A critical global assessment of atomic energy.* World Scientific; Mark Diesendorf (2014) *Sustainable Energy Solutions for Climate Change.* Routledge and UNSW Press, chapter 6; Schneider & Froggatt (2021), *op. cit.*

35. Anon (2022). *Power.* https://www.powermag.com/blog/former-nuclear-leaders-say-no-to-new-reactors

36. Jan Willem Storm van Leeuwen (1985). Nuclear uncertainties: energy loans for fission power. *Energy Policy* 13(3), 253–266; Manfred Lenzen (2008). Life cycle energy and greenhouse gas emissions of nuclear energy: a review. *Energy Conversion & Management* 49, 2178–2199.

37. ITER. https://www.iter.org/; critiqued by Daniel Jassby (2018). https://thebulletin.org/2018/02/iter-is-a-showcase-for-the-drawbacks-of-fusion-energy

38. For more detailed discussions on renewable energy policies, see Mark Diesendorf (2014). *Sustainable Energy Solutions for Climate Change.* Routledge and UNSW Press; IRENA, IEA and REN21 (2018). *Renewable Energy Policies in a Time of Transition.* https://www.irena.org/publications/2018/apr/renewable-energy-policies-in-a-time-of-transition; Scott Valentine, Marilyn Brown & Benjamin Sovacool (2019). *Empowering the Great Energy Transition: Policy for a low-carbon future.* Columbia University Press.

39. AEMO (2020). *Integrated System Plan 2020*. Australian Energy Market Operator. https://aemo.com.au/-/media/files/major-publications/isp/2020/final-2020-integrated-system-plan.pdf?la=en&hash=6BCC72F9535B8E5715216F8ECDB4451C

40. IISD (2020). *Doubling Back and Doubling Down: G20 scorecard on fossil fuel funding*. https://www.iisd.org/system/files/2020-11/g20-scorecard-report.pdf

41. IEA (n.d.). *Energy subsidies*. https://www.iea.org/topics/energy-subsidies

42. Tim Laing et al. (2013). *Assessing the Effectiveness of the EU Emissions Trading Scheme*. Grantham Research Institute. https://www.jcfj.ie/wp-content/uploads/2016/02/WP106-effectiveness-eu-emissions-trading-system.pdf; Easwaran Narassimhan et al. (2018). Carbon pricing in practice: a review of existing emissions trading systems. *Climate Policy* 18(8):967–991. https://doi.org/10.1080/14693062.2018.1467827

43. S&P Global Commodity Insights. https://www.spglobal.com/platts/en/market-insights/latest-news/coal/120120-germany-awards-coal-closure-compensation-to-48-gw-to-shut-2021

44. Frank Jotzo & Salim Mazouz (2015). *The Conversation*. https://theconversation.com/farewell-to-brown-coal-without-tears-how-to-shut-high-emitting-power-stations-50904/

45. Eric Campbell (2021). *ABC News*. https://www.abc.net.au/news/2021-09-02/the-just-transition-as-spain-says-goodbye-to-coal/100421216

46. The standard frequency of alternating current is 50 Hz (Herz = cycles per second) in most of the world and 60 Hz in the USA.

47. For example, Reposit Power. https://repositpower.com

48. Jacobson & Delucchi (2015) *op. cit.*; Mark Diesendorf & Ben Elliston (2018). The feasibility of 100% renewable electricity systems: a response to critics. *Renewable & Sustainable Energy Reviews* 93:318–333. https://doi.org/10.1016/j.rser.2018.05.042

49. Ben Elliston, Jenny Riesz & Iain MacGill (2016). What cost for more renewables? The incremental cost of renewable generation—An Australian National Electricity Market case study. *Renewable Energy* 95:127–139. Blakers et al. (2017), *op. cit.*

50. Dimitrii Bogdanov et al. (2021). Low-cost renewable electricity as the key driver of the global energy transition towards sustainability. *Energy* 227:120467. https://doi.org/10.1016/j.energy.2021.120467

51. AEMO (2021). *Application of Advanced Grid-scale Inverters in the NEM*.   https://aemo.com.au/-/media/files/initiatives/engineering-framework/2021/application-of-advanced-grid-scale-inverters-in-the-nem.pdf?la=en

52. Not to be confused with synchronous generators.

53. Giles Parkinson (2021). *Renew Economy*, 8 August. https://reneweconomy.com.au/wind-and-solar-to-get-taste-of-freedom-as-new-syncons-join-the-grid/

54. Giles Parkinson (2021). *Renew Economy*, 5 August. https://reneweconomy.com.au/aemo-to-fast-track-grid-forming-inverters-to-help-transition-to-100-renewables/

55. Giles Parkinson (2016). *RenewEconomy*, 1 November. https://reneweconomy.com.au/oz-minerals-looks-to-solar-to-help-power-1-billion-copper-project-84367/Giles   Parkinson   (2021).   *RenewEconomy*,   2 February.   https://reneweconomy.com.au/bhp-signs-deal-with-w-a-s-biggest-solar-farm-to-supply-half-of-refinerys-power-needs/   Giles Parkinson (2021). *RenewEconomy*, 7 September. https://reneweconomy.com.au/potash-mine-to-build-wind-solar-and-battery-micro-grid-for-most-of-its-power-needs/

56. Giles Parkinson (2021). *RenewEconomy*, 29 October. https://reneweconomy.com.au/bluescope-and-rio-in-renewable-steel-venture-fortescue-buys-green-hydrogen-tech/

57. Joshua Hill (2021) *RenewEconomy*, 19 January. https://reneweconomy.com.au/dubai-solar-park-says-it-has-started-powering-aluminium-production/

58. RE100. https://www.there100.org/re100-members

59. Net energy is energy output minus life-cycle primary energy input. Energy return on energy invested (EROEI or EROI) is defined to be energy output divided by life-cycle primary energy invested. The life-cycle spans mining the raw materials, transporting and processing them, constructing the technology and dismantling it at the end of its working life.

60. Enrica Leccisi, Marco Raugei & Vasilis Fthenakis (2016) The energy and environmental performance of ground-mounted photovoltaic systems–a timely update. *Energies* 9:622.

61. Raugei M, Leccisi E (2016) A comprehensive assessment of the energy performance of the full range of electricity generation technologies deployed in the United Kingdom. *Energy Policy* 90:46–59.

62. Brockway PE, Owen A, Brand-Correa LI et al. (2019) Estimation of global final-stage energy-return-on-investment for fossil fuels with comparison to renewable energy sources. *Nature Energy* 4:612–621

63. Mark Diesendorf & Thomas Wiedmann (2020). Implications of trends in energy return on energy invested (EROI) for transitioning to renewable electricity. *Ecological Economics* 176:106726. https://doi.org/10.1016/j.ecolecon.2020.106726

64. This is sometimes called the dynamic EROI issue, while EROIs of individual technologies at a particular time have static EROIs (Diesendorf & Wiedmann 2020, *op. cit.*, Section 3.3). Note: in both the 3rd line and 8th last line of Section 3.3 of that paper, 'temporary increase' should be replaced by 'temporary decrease'.

65. Paul Denholm et al. (2009). Land-use requirements of modern wind power plants in the United States. Technical report NREL/TP-6A2-45,834. National Renewable Energy Laboratory, Golden CO.

66. Rémi Rauline et al. (2021). Sharing the sky. *Renew*, issue 154, 60–64.

67. Bent Sørensen & Peter Meibom (2000). A global renewable energy scenario. *Int. J. Global Energy Issues* 13:196–276; Mark Jacobson et al. (2017a) 100% wind, water, and sunlight all-sector energy roadmaps for 139 countries of the world. *Joule*, 1, 108–121. https://doi.org/10.1016/j.joule.2017.07.005

68. Jørgen Nørgaard (2008). Avoiding rebound through a steady-state economy. In: H. Herring & S. Sorrell (Eds.), *Energy Efficiency and Sustainable Consumption: The Rebound Effect*. Palgrave Macmillan.

69. David Font Vivanco et al. (2016). How to deal with the rebound effect? A policy-oriented approach. *Energy Policy*, 94:114–125. https://doi.org/10.1016/j.enpol.2016.03.054

70. Tobias Junne et al. (2020). Critical materials in global low-carbon energy scenarios: The case for neodymium, dysprosium, lithium, and cobalt. *Energy 211*, 118,532. https://doi.org/10.1016/j.energy.2020.118532

71. An extraordinary article by Megan Seibert and William Rees (2021), *Energies* 14 (15), 4508, https://doi.org/10.3390/en14154508, claims that we must return to a society without modern wind and solar power where homes are heated by firewood and transport is by horse and buggy. This article was critiqued severely in independent articles by Vasilis Fthenakis et al. (2022) Energies 15(3), 971, https://doi.org/10.3390/en15030971 and Mark Diesendorf (2022) *Energies* 15(3), 964, https://doi.org/10.3390/en15030964. The editorial in *Energies* 15(3), 889;

https://doi.org/10.3390/en15030889 stated that, "the original Seibert and Rees manuscript...slipped through our system in spite of the warning signals given by two of our reviewers...[I ask] our readers and our constituency to forgive me for accepting the original manuscript without requiring the authors to make some obvious corrections (that, in light of their response reported below, I believe they would not have accepted)". The replies to their critics by Seibert and Rees are mainly generalities (e.g. about 'overshoot') that avoid the specific points made by Fthenakis et al. and by Diesendorf.

72. IEA (2021b), *op. cit.* The problem is discussed in more detail by Mark Diesendorf & Steven Hail (2022). Funding of the energy transition by monetary sovereign countries. *Energies* 15:5908. https://doi.org/10.3390/en15165908

73. Diesendorf & Hail (2022) *op. cit.*

74. See for example, Scenarios 14 and 16 in Mark Diesendorf (2022a). Scenarios for mitigating CO_2 emissions from energy supply in the absence of CO_2 removal. *Climate Policy*, https://doi.org/10.1080/14693062.2022.2061407

Reference

Weblinks accessed 26/10/2022.

5

Transitioning Natural Resources

The truth is: the natural world is changing. And we are totally dependent on that world. It provides our food, water and air. It is the most precious thing we have and we need to defend it. (David Attenborough [1])

On the edge of Jakarta stands a pile of garbage more than 15 stories high, spanning 200 football fields. Each day it grows by another 6500 tonnes. While Bantar Gebang[2] is one of the largest landfill sites in the world, it is emblematic of the linear or one-way economy in which resources are extracted, used, then dumped. In extreme cases the useful life of the resources can be measured in minutes—an item is purchased in a plastic bag that is almost immediately discarded, is blown into a drain from where it enters a river and ends up in the ocean and perhaps is ingested by a fish and then a human, or adds to one of the great ocean garbage patches. In the 'best' case, it ends up in landfill.

These results are largely a product of the present economic system: for many products, it is far cheaper to extract resources such as timber or minerals, feed them into a manufacturing process and, when they have lost value, throw them away. The word 'waste' neatly encapsulates this situation: the verb means 'to use, consume, spend, or expend thoughtlessly or carelessly'. Yet landfill represents a vast reserve of resources that could be used instead of buried.

© The Author(s), under exclusive license to Springer Nature Singapore Pte Ltd. 2023
M. Diesendorf, R. Taylor, *The Path to a Sustainable Civilisation*,
https://doi.org/10.1007/978-981-99-0663-5_5

Indeed, this is what the people of Bantar Gebang are doing when they pick through the piles looking for any items of value. While for them it's a way of dealing with poverty, a far better solution would be that the economic system did not leave them behind and 'waste' was reduced, reused, remanufactured and recycled rather than tipped into ever growing piles. Instead of 'cradle to grave', we must think about the life cycle of resources in terms of 'cradle to cradle'[3] or, to use a more recent term, a circular economy instead of a linear one.[4] A simple analysis of the materials that end up in landfill suggests the vast untapped potential of 'wastes'.[5]

In general, it would save energy, other resources and money to address natural resource conservation in the following order of priorities: reuse, remanufacture, recycle and dispose of the remainder.[6] Even before dealing with the fate of product, it would make sense for governments to legislate that products must be designed for easy repair, reuse, remanufacture and recycling. However, industry resists any such attempts with arguments that it would damage the economy (i.e. their profits) and by shifting the responsibility from the manufacturer onto the consumer, with such campaigns as Keep America Beautiful.[7]

To discuss the potential for making our use of natural resources more sustainable, it is convenient to divide them into renewable and non-renewable resources.

5.1   Renewable Resources

A necessary, but not sufficient, condition for sustainability of renewable resources is to ensure that their rates of use do not exceed their rates of regeneration—easy in theory but difficult in practice. This section outlines the challenges of transitioning agriculture, food and diet—and managing soils, forestry, freshwater and fisheries—for the transition to Sustainable Civilisation.

Agriculture and Food Supply

We as consumers are disconnected from our food sources. Food arrives in supermarkets, usually wrapped in plastic. We don't know whether it is local or from the other side of the world, unless we read the labels carefully. We generally don't know whether it's grown causing environmental damage or exploited labour. This disconnection both reflects civilisation's greatest strength and its greatest weakness. Food and other products appear as if by magic and our isolation from their origins encourages us to ignore their impacts. We don't think about it until supply is threatened.

One of the core foundations of the Sustainable Civilisation, therefore, is to rethink the resilience of our food supply systems and, in particular, agriculture. While the consumer disconnect with the origin of products is probably an inevitable consequence of complex supply chains, there is much that can be done to improve the sustainability of agriculture.

Before discussing that, we should point out that, while the above supermarket experience relates to wealthy nations, for others obtaining food is already a stark reality. In April 2022, Director-General of the World Trade Organisation, Ngozi Iweala, issued a blunt warning relating to Putin's war in Ukraine[8]: "*I think we should be very worried,*" she said. "*The impact on food prices and hunger this year and next could be substantial. Food and energy are the two biggest items in the consumption baskets of poor people all over the world.*" It's a situation that is affecting people in many other countries, since Ukraine would normally export huge quantities of wheat and Russia is a major manufacturer of fertiliser. Food shortages could destabilise the Middle East, in particular Egypt, the most populated Arab country, which imports 80% of its wheat from Ukraine and Russia.[9]

As the world plunges into climate change, our ability to grow sufficient food is becoming more challenging. While agriculture has always faced the vagaries of weather, the relative climate stability of the last 10,000 years is disappearing and formerly productive areas are under threat. The environmental challenge is to maintain food production while not increasing its land use.

One solution is regenerative agriculture, discussed below under Soils. Science writer Julian Cribb explores the nexus between food availability and conflict; his solutions include recycling nutrients and wastewater, and expanding food production in cities.[10]

Synthetic Meat

Synthetic, cultured or lab-grown meat is at an early stage of development. It is produced by taking a small quantity of stem cells from a living animal and placing them in a serum in a sterile bioreactor where they grow and multiply. Its potential advantages include a huge reduction in land use, reduced pollution of rivers by agricultural chemicals, rapid growth compared to cattle and sheep, improved animal welfare, the absence of antibiotics in the 'meat', and the ability to control the fat content and to replace saturated fats with omega-3 fatty acids. Concerns and potential disadvantages include its very high cost (which will decline with mass production), the risk of contamination if an animal-based serum is used, the likely use of growth hormones in the serum, and possible long-term safety issues. There is debate about the magnitudes of the life-cycle energy inputs, greenhouse gas (GHG) emissions and water use of the processes. Overall, some sustainable, organic, open pasture livestock farms may actually be better environmentally than lab-grown meat. The jury is still out.

Diet, Health and the Environment

What we eat affects both the health of the planet and our own health. Food production has one of the greatest combinations of environmental impacts of all economic activities. Methane emissions from cattle and sheep, and nitrous oxide emissions from fertiliser use, account for about 11% of total global anthropogenic emissions of greenhouse gases.[11] Agriculture is also responsible for biodiversity loss, freshwater use, interference with the global phosphorus and nitrogen cycles, land use change and chemical pollution. At the very least, ecologically sustainable food production should result in net zero GHG emissions, reduce land use and have fewer other impacts.

Although there are in theory sufficient calories for the world's population, they are not distributed equally. Furthermore, many people, in both rich and poor countries, have nutritional deficiencies resulting in such diseases as obesity, coronary heart disease, stroke, diabetes, some cancers, rickets, osteomalacia and diarrhoea. The interdisciplinary EAT-Lancet Commission identified a healthy reference diet as an alternative to present diets for estimating health and environmental impacts[12]:

> This healthy reference diet largely consists of vegetables, fruits, whole grains, legumes, nuts, and unsaturated oils, includes a low to moderate amount of seafood and poultry, and includes no or a low quantity of red meat, processed meat, added sugar, refined grains, and starchy vegetables.

The broad features of this diet are compatible with many cultures around the world. The EAT-Lancet Commission found that it could provide for about 10 billion people in 2050 and remain within the safe operating space of the environment. *"However, even small increases in the consumption of red meat or dairy foods could make this goal difficult or impossible to achieve."*[13]

Biodiversity

Although we humans imagine that we are the lords and ladies of creation, we are actually completely dependent upon the natural world for our survival, as emphasised in Chap. 2. In the living world, every organism is connected to every other species in a complex web of life. Conserving biodiversity (defined in Box 5.1) is vital.

Box 5.1 Biological Diversity

Biological diversity (biodiversity) comprises the millions of different species that live on our planet—plants, animals, and microorganisms—as well as genetic differences within each species—and all the different ecosystems, in which groups of particular species form a unique community, interacting with one another and with the air, water, and soil around them.

It's possible that we could survive without mosquitos and blowflies, but it's very unlikely that we could survive if all mycorrhizal fungi died as the result of, say, chemical pollution (see Box 5.2). Similarly, without bees to pollinate plants, there would be no fruit and nuts. Bee populations are declining under pressure from pesticides, habitat loss, climate change, diseases and parasites such as the Varroa mite.

Box 5.2 Mycorrhizal Fungi

Mycorrhizal fungi extract nutrients, including phosphorus, from the soil, making them available to the tree roots to which they are attached. The tree converts these nutrients from inorganic to organic forms that it needs and, in return for the fungi's service, makes carbon available to the fungi. Mycorrhizal fungi have evolved vast underground networks, which exchange nutrients, chemicals and even electrical signals between the roots of diverse host plants.[14] The operation of this symbiosis or mutualistic association is vital for the Earth's carbon balance and the health of the vast majority of trees. Trees are essential for maintaining our atmosphere, climate and much more.

Human activities are killing species at a thousand times the natural rate.[15] We do not know which of these species are essential for our survival or at least are important for our thriving.

In addition to this instrumental reason for the conservation of biodiversity are the ethical arguments that it should be conserved for its scientific, social, cultural, spiritual, educational, recreational and aesthetic values.

Biodiversity can be conserved by expanding national parks and other nature reserves; legislating to protect particular endangered species; mitigating climate change; reducing the prevalence of invasive species; restoring habitats (including forests, wetlands and waterways); captive breeding of endangered species; seed banks; eco-labelling; research; and education. However, the greatest threat to biodiversity is the destruction of habitats through agriculture and urban sprawl—major reforms are needed in both these activities.

Soil

Without soil, there would be almost no food. Far from being inanimate dirt, soil is a rich ecosystem with a complex chemical and biological structure. Aside from gross removal by erosion caused by the loss of ground cover, soils have been damaged through being treated as a replaceable industrial asset. Overexploitation and inappropriate chemical use have stripped the life out of many soils, depriving them of their sustaining insects, fungi, bacteria and essential nutrients. Soils are a vital part of the climate change story. Over the past 12,000 years, disturbed natural ecosystems like forests and grasslands have released vast amounts of stored soil carbon into the atmosphere. Specifically, in the USA, the growth of farmland has released about 110 billion tonnes of carbon from the top layer of soil[16]—roughly equivalent to 80 years' worth of present-day US emissions.[17]

If agriculture is to remain viable, soil needs to be nurtured as a living organism. The science on this has advanced considerably over recent decades, with varying approaches called 'regenerative agriculture', 'integrated agriculture', 'organic farming', and so on.[18] 'Regenerative agriculture' describes agricultural practices that focus on the health of the ecological system as a whole, not solely on high production yields of crops. Although it lacks a formal definition, it is generally agreed that its main principles are improving soil health, increasing biodiversity, aiding in carbon sequestration, incorporating humane treatment of livestock and farmworkers, and improving the overall larger ecosystem as a whole.[19]

Freshwater

If soil is one pillar of agriculture, freshwater is another. Unfortunately, humanity has a poor record in conserving rivers and wetlands. Wetlands once covered around 10% of the earth's land surface but, in only 50 years, half the world's mangrove forests have vanished. Wetlands also occupy prime waterfront real estate where we build houses, hotels, marinas and dockyards. Wetlands are very effective at capturing CO_2.[20] The remaining mangroves that currently cover 14–15 million hectares around the world trap an estimated 31–34 million tonnes of carbon every year.[21]

The Global Database of Dams [22] has catalogued about 34,000 dams. As the effects of climate change and population growth bite, there is pressure to build more, both for drought-proofing water supplies and to mitigate floods. Each dam affects a river in various ways, including[23]:

- deepening riverbeds and hence lowering groundwater, thus reducing its access to plant roots and to human communities dependent on wells;
- inundating upstream land and so modifying its temperature, chemical and physical properties, thus enabling invasive species;
- altering downstream water flow and sediment transport.

In places where ground cover has been removed, erosion washes away topsoil and nutrients, which then flow into rivers and streams. Altered water flows damage stream beds, riparian zones and aquatic life. In the Amazon basin and parts of Asia, large-scale land clearing is producing long-term damage to soils and hydrological systems.[24] In urban environments, waterways have been treated within a classic engineering mindset: water is a problem, a thing to be removed, and natural waterways have been replaced by concrete drains.

The water situation affects more than agriculture, because changes to the hydrological cycle can lead to drought and accompanying wildfires. The drying climate, combined with ever greater demands on water, has resulted in the Californian food-bowl now facing possibly a permanent water shortage crisis.

Fortunately, land-owners are increasingly aware that management of their water resources involves working with nature, not against it. Water is best retained in the landscape and in the soils where it is of greatest benefit. Irrigation practices can be better tuned to soil moisture conditions and applied in ways that minimise losses through evaporation. Like soil, rivers and their riparian zones are important carbon sinks. It is estimated that a degraded river stores less than 2% of the carbon of a healthy one.[25]

Fisheries

Across the planet, fish are a crucial source of human nutrition, supplying more than 2.6 billion people with at least 20% of their average animal protein intake.[26]

Harvesting of fish resources has increased markedly, and it's now estimated that 90% of all big fish have been taken from the oceans.[27] Removal of large fish, such as tuna, cod and groupers, has profound knock-on impacts through the marine environment. As in the case of chopping down old growth forests, the larger, older members play critical role in the ecosystem. Big fish produce the most offspring, so that catching them depletes the next generation.

Now, super trawlers the size of a football pitch cross the oceans, scooping up everything into enormous nets. By one estimate, it would take 56 traditional African fishing boats a year to harvest an equivalent haul in a single day[28] and it's predicted that practices such as these will lead to the collapse of all seafood fisheries by 2050. A 2016 study[29] estimates that more than 10% of the world's population will be at risk of a deficiency in key nutrients if fish stocks continue to decline.

However, as the UN Global Compact states,[30] there are huge opportunities: "*The ocean is a pillar of global food security with a critical role to play in supplying a growing global population with sustainable protein.*" Countries such as the USA, Britain and Australia have established quotas and no-fish zones that provide sanctuary breeding habitats.

When poorly done, fish farming in coastal waters causes considerable damage to ecosystems. The intense concentration of fish in a small area injects an abnormal nutrient load into the water, depleting oxygen levels and triggering algal blooms. Farmed fish may contain antibiotics that produce resistant strains of bacteria in people who eat them.[31] Growing a kilo of salmon requires at least three kilos of wild-caught fish. Fortunately, it is possible to replace some of this using plant ingredients. In Norway, for example, fishmeal and fish oil has dropped to around 30% from 90% of salmon feed in the 1990s.[32]

The approach of 'regenerative aquaculture' places human harvesting of marine life within the context of an ecology where ecosystem services do much of the work for free. In one example,[33] aquaculture incorporates fish, seaweed and crustacea in an integrated system. Fish effluent provides nutrients for kelp and shellfish, which in turn help feed the fish. Seaweed is a highly productive crop containing high amounts of polyunsaturated (omega-3) fatty acids, potassium, iron, calcium and fibre. The cited study claims that a typical 20-acre (8-hectare) farm can produce 60 tonnes of kelp and 250,000 shellfish per year while sequestering nine tonnes of carbon dioxide and 300 kg of nitrogen.

Forests

We depend on forests for our survival and our ability to thrive. They mitigate climate change by absorbing CO_2 from the atmosphere and in turn emit oxygen, essential to life on Earth. They regulate water flows and prevent soil erosion. They are homes for indigenous people and for many species, from the animal kingdom to fungi, plants and many microorganisms. They are complex ecosystems that host a substantial part of our planet's biodiversity and store genetic resources. They are beautiful, relaxing places for increasing our health and wellbeing by hiking, running and meditating.

Yet, despite our dependence on forests for vital ecosystem and health services, we are allowing them to disappear.[34] About 40% of our planet's primary forests have been destroyed to produce food and other products desired by industrial society—timber, paper and ingredients of medicines, cosmetics and detergents—and to make way for sprawling towns and suburbs. We continue to lose about 10 million hectares annually,[35] mostly from the growth of agriculture. The destruction of the Amazon rainforest is discussed in Box 5.3.

Box 5.3 Destruction of the Amazon Rainforest

The world's largest tropical rainforest is on fire. Satellite data recorded over 70,000 fires in Brazil between January and August 2022, with more than half in the Amazon region. Most of these fires are started deliberately, by farmers clearing land primarily for cattle ranching, but also for crop production; by miners seeking gold, copper, iron, manganese, etc.; and by the forestry industry. About one-quarter of the Amazon rainforest has already been lost or badly degraded.[36] In addition, pesticides, fertiliser and mercury (from gold mining) pollute the soil and waterways.[37]

The Amazon rainforest used to be a carbon sink, absorbing CO_2 from the atmosphere and giving us oxygen in return. But as the forest burns, it releases the CO_2 back into the atmosphere— south-eastern Amazonia has already become a net carbon source.[38] Deforestation and burning are destroying the home of the Indigenous people, who have lived there possibly for thousands of years, and are decimating the biodiversity of the most biodiverse region of the world.

These impacts, which have been occurring for decades, accelerated following the election of Jair Bolsonaro as president in 2019. Environmental protection policies were revoked, deforestation law enforcement was halted, and the rate of deforestation increased to over 10,000 square kilometres annually. Research published in 2021 estimates that 94% of deforestation in the Amazon and neighbouring Cerrado regions could be illegal.[39] Replacing Bolsonaro would probably reduce the rate of deforestation temporarily, but would not solve the problem. Land grabbing is widespread, especially in 'non-designated' land (land whose legal status has not been determined).[40] Land tenure rules must be enforced, the occupation of non-designated land punished and the protection of traditional and Indigenous populations strengthened.[41]

Forests are being destroyed because their environmental and social values are being sacrificed to the demands of economic and population growth, within the dominant economic system which depends on exploitation of the environment and working people's labour (see Chap. 7). Although some countries have laws that nominally restrict logging and land clearing, much of the ongoing devastation is conducted illegally.

Deforestation has become an important cause and effect of climate change. Global heating is increasing the incidence and severity of forest fires and with them the loss of biodiversity and the growing danger of more extinctions. In recent years, wildfires have burned vast areas of forests in Siberia, Europe, North Africa, North America, the Amazon region[42] and Australia (see Box 5.4). The higher temperatures have also stimulated proliferation of the mountain pine beetle, which is decimating the pine forests of Canada, turning them into fuel for wildfires.[43]

Box 5.4 Australia's 'Black Summer' Bushfires

The 'Black Summer'[44] bushfires in 2019–2020 in Australia burned over 24 million hectares of land, destroyed over 3000 homes and took 33 lives, including six Australian and three American firefighters. In addition, hundreds of people may die prematurely from the inhalation of smoke that blanketed vast regions. The fires killed or displaced an estimated three billion animals, including some rare or threatened animals, and many plant and insect species. The economic impacts amounted to over AUD 10 billion.[45]

Total GHG emissions from the conflagration, estimated at 830 million tonnes,[46] were much greater than Australia's official annual territorial emissions of 488 million tonnes in 2021.[47] International accounting of GHG emissions excludes emissions from wildfires on the questionable assumption that absorption of CO_2 from regrowth balances emissions from consumption. However, in practice some burnt areas regrow slowly, while others never recover.

The solutions we seek must simultaneously conserve forest ecosystems, increase human wellbeing and preferably generate income for people living in and near forests. Agriculture and forests can coexist. The Food and Agriculture Organisation of the United Nations recommends a strategy comprising

> three forest and tree-based pathways that complement other actions aimed at achieving more efficient, more inclusive, more resilient and more sustainable agrifood systems: namely: halting deforestation and maintaining forests; restoring degraded lands and expanding agroforestry; and sustainably using forests and building green value chains.[48]

Consumer and environmental NGOs must campaign for legislation to ensure that all timber sold in their jurisdictions is certified by the Forest Stewardship Council.[49]

Renovating towns and cities to reduce (sub)urban sprawl, by concentrating population around public transport nodes,[50] reduces energy use in transportation, air pollution and GHG emissions, improves lifestyles of low-income people currently forced to the fringes of expensive cities, and allows biodiversity to increase near towns, while reducing land clearing and deforestation.

Thus, national and international action to mitigate human-induced climate change can reduce the destruction of forests.[51]

Recommendations

Unfortunately, the sustainable solutions to the decline of renewable natural resources are, in many cases, the opposite of current trends. We must change agricultural practices in order to regenerate soils, waterways and ecosystems. We must change diets to the healthy reference diet with low meat in order to reduce land use, GHG emissions and other environmental impacts of agriculture while improving human health and wellbeing. Expansion of fish quotas and marine protected areas will actually increase the supply of fish. Continuing research and development is needed on synthetic meat and its environmental and health impacts. To protect soils and biodiversity, deforestation must be ended, reforestation and agroforestry expanded, and urban sprawl reduced. Consumers must demand better labelling of food and sources of timber.

Campaigns by environmental and consumer organisations on all these issues have slowed destruction but not stopped it. Therefore, the campaigns must be expanded in scope to weaken the fundamental driving forces, namely vested interests (Chap. 6) and an inappropriate economic system (Chap. 7). Simultaneously, social justice must be improved and democratic decision-making strengthened.

5.2   Non-renewable Resources

Imagine you are flying in a hang glider, soaring high above the ground when you look up and notice that one crucial bolt is breaking and, when it fails, you will crash. The Jesus Bolt gets its name from the imagined last words uttered by someone who has discovered a single point of failure. There are a number of Jesus Bolts in our civilisation, some of which are more obvious than others. Consider, for example, what would happen if the internet failed. With no means to transmit money, the economy would collapse and, with it, society. In this case, the crucial role of the internet is well understood and maintaining it is given considerable attention. Fortunately, that makes the internet fairly resilient, although a dramatic television series has been created around an extreme solar event that brings down global communications systems and more.[52] However, the exhaustion of several key minerals and technologies would be a different story: these are the Jesus Bolts of an ecologically sustainable civilisation. This section discusses a few such resources that are likely to fail unless we take urgent action.

Substituting Renewable for Non-renewable Resources

Non-renewable resources can be difficult to conserve in an ecologically sustainable society unless they can be replaced with renewable resources. Where this is possible, it is essential to find substitutes that are environmentally and socially sustainable and affordable.

Chapter 4 shows that we have a clear pathway to the replacement of fossil fuels used for energy production with renewable energy, mostly wind and solar. However, substituting renewable resources for the *non-energy* industrial uses of fossil fuels is more difficult, still requiring research and development to reduce the costs of producing 'green' hydrogen and ammonia.

Cross-laminated timber (CLT)[53] is a renewable resource, provided the timber is obtained from sources certified by the Forest Stewardship Council or, even better, reforestation. As a construction material, it has much lower embodied energy than the non-renewable construction materials, bricks, conventional concrete and steel. It is being increasingly used in small buildings on account of its structural strength, attractive appearance, rapid on-site construction and lesser waste. Even skyscrapers are being built with CLT components.

Irreplaceable, Essential, Non-renewable Resources

Next, we consider the other extreme: irreplaceable but essential non-renewable resources. Phosphorus merits special attention, since it is specifically included as a dimension in the Planetary Boundaries framework discussed in Chap. 1. Phosphorus entering the oceans as a result of human activities amounts to more than ten times the natural background, pushing it outside Earth's safe operating zone.[54]

Phosphorus is an essential, irreplaceable mineral for humans and, indeed, all life forms. It is a component of DNA, RNA, many proteins, bones, teeth, cell membranes and of the body's key energy source, ATP. It plays key roles in regulation of gene transcription, activation of enzymes, maintenance of normal pH in extracellular fluid, and intracellular energy storage.[55] Without it, global food production would plummet.[56]

Tragically, huge amounts are sent into sewers every day. Of the domestic contributions to the estimated 44,000 tonnes of phosphorus that enter United Kingdom sewage treatment works each year, the main contributions are from natural diet (40%), food additives (29%), laundry (14%) and detergents (9%)[57] Similar discharges are expected from other developed countries. It is estimated that source controls could halve these discharges to sewerage works and hence to the aquatic environment.[58]

Its major use and the major cause of its loss into the aquatic environment are in agriculture, through the inefficient use of phosphate fertilisers. Field experiments find that often only 10–15% of the phosphorus applied to a crop is taken up by it. Yet once the phosphorus level in soil has been built up to a critical level, further applications to the soil can reach efficiencies of 80–90%.[59] Extending these efficiencies across the world could increase the lifetime of the principal source of phosphorus, phosphate rock, by centuries, buying us time to develop low-cost, low-energy methods to reduce the wastage of phosphorus and even possibly recycling it. Reducing the discharge of phosphorus into freshwater creeks has an environmental benefit: it reduces the toxic algal blooms and excessive plant growth that damage biodiversity and water quality for domestic use, fish stocks and the recreational use.

Scarce Non-renewable Resources in General

The potential scarcity of certain non-renewable resources can be managed to some degree by asking a series of questions of ourselves and the people in power (see Fig. 5.1) and then acting upon them. The first question is: Do we, both as individuals and as a society, really need the products made from this scarce resource? After all, a good life is still possible without many of the gadgets and chemicals on the market in the twenty-first century. If a product's life cycle is very damaging to the health of people and the environment, then its production and use can be controlled or eliminated by taxation or regulation. If some products made from this resource are indeed vital, then our second question is: Can we substitute products that use less or none of the scarce resource or, alternatively, products that use a more common resource? If the answer is 'no', demand can be reduced by improving the efficiency of the production and use of the products. This buys us time for answering the third question: How can we reuse and, failing that, recycle the scarce, essential, irreplaceable raw materials?

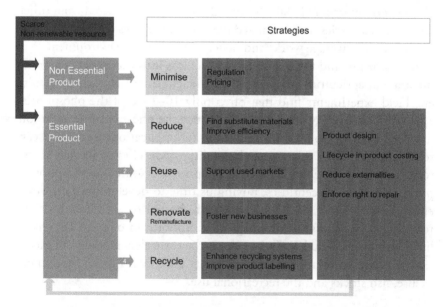

Fig. 5.1 Questions and possible answers about non-renewable resources

In the case of phosphorus, the last resort would be attempting to extract phosphorus from freshwater creeks and rivers and the ocean. Phosphorus is dispersed there at very low densities and so, unless a simple chemical or biological process can be discovered to bind and concentrate it, an enormous quantity of energy would have to be expended to recover it. It's much more fruitful to extract it at sewerage works[60], but this does not capture direct losses from farms into the aquatic environment. Another obstacle is the present economic system, which does not value phosphorus as being essential and irreplaceable.

Fossil fuels, especially oil and gas, have many uses outside the energy sector; see Table 5.1. Many of these do not involve the emissions of GHGs and some that do can be substituted with renewable materials.

Essential Minerals for Renewable Energy Technologies

It would take an entire volume to comprehensively examine non-renewable raw materials, so instead we discuss briefly several of the key elements required for the transition to renewable energy.

The raw material for almost all solar photovoltaic cells is silicon, one of the most abundant elements on the planet. Silicon is processed to a high-purity, polycrystalline form called polysilicon. The production of polysilicon fluctuates with demand. During the COVID-19 pandemic, its production decreased and so its price increased sharply. Subsequently, several new factories commenced construction and so the availability of polysilicon is expected to increase (and its price to decrease) in 2023. The problem is not any shortage of silicon but rather the need to plan to meet the expected future rapid increase in demand for polysilicon.

Table 5.1 Non-energy uses of fossil fuels

Steel-making	Aluminium manufacturing	Lubricants	Paints & coatings
Wax	(Brown) hydrogen	Synthetic fibres	Asphalt
Ink & dyes	Pharmaceuticals	Adhesives	Elastomers
Plastics	Insecticides	Herbicides	Polystyrene
Cosmetics	Soaps & detergents	Fertiliser	Vitamins
Explosives	Industrial chemicals	Resins	Foams

Source: The authors' survey of the literature

There is a booming demand for lithium to make batteries for electric vehicles (EVs), electrical devices and storage of renewable electricity for homes, industries and electricity grids. Lithium batteries have several advantages compared with other batteries: high energy density; low maintenance; and uses ranging from small electronic devices to grid-scale electricity storage. Their disadvantages include the adverse environmental and social impacts of lithium mining,[61] hazardous materials, short lifespan, standing losses and high cost of lithium batteries for storage of more than several hours. While there are abundant reserves of lithium to meet short-term demand, in the absence of substantial recycling and the development of alternatives, the medium- and long-term sustainability of transportation based on EVs is at risk.[62] For example, it's expected that, "by 2030, the European Union will need up to 18 times more lithium and up to five times more cobalt, compared with the [2020] supply".[63] The recycling of lithium-ion batteries has only commenced recently.[64] Substitute batteries based on more common materials are under development and a few (e.g. flow batteries) are already in use,[65] but so far alternatives cannot match lithium's high energy density and so are less suitable for EVs and many consumer electronic devices such as mobile phones.

The rare earth elements neodymium and dysprosium are used in electric motors/generators in EVs and wind turbines, while cobalt is used in some types of lithium batteries. Together with lithium, they are all subject to potential supply bottlenecks and so a risk exists that, as the energy transition accelerates, future demand could exceed reserves.[66] The problem is potentially soluble by design for disassembly, recycling, substitution, and improved efficiency of manufacturing (and, of course, by reducing global energy consumption), but increased action is needed.

Clearly, the 'free' market cannot be relied upon to deliver the rapidly increasing demand for essential minerals needed by the energy transition. Interventions by national governments and international agreements are needed to guide the market. Policy options include caps on production of scarce non-renewable resources; resource taxes with revenue distributed as dividends to all adult citizens; environmental certification; and regulations with penalties to discourage planned obsolescence, unnecessary packaging and products that cannot be reused or recycled.

5.3 Green and Appropriate Technologies

Australia's former Prime Minister Scott Morrison used to dismiss public concerns that the nation's climate mitigation efforts were inadequate with the slogan, "Technology not taxes". He claimed that, "World history teaches one thing—technology changes everything." But is technology enough? One of the themes of this book is that a single, simplistic approach to achieving sustainability will not succeed. While there is no doubt that technology is necessary, it is far from sufficient. Indeed, that has been a difficult lesson for us authors since we both have long professional careers in technology (MD in energy; RT in IT systems). Solutions rarely fail due to technical problems. Other dimensions to the solution are considered elsewhere in this book. In this section we consider what appropriate 'green' technologies look like.

But first, what do we mean by 'technology'? A conventional definition of 'technology' and one of its shortcomings are outlined in Box 5.5.

Box 5.5 What Is Technology?

According to Encyclopaedia Britannica: *"Technology is the application of scientific knowledge to the practical aims of human life or, as it is sometimes phrased, to the change and manipulation of the human environment."*[67] This definition is unsatisfactory because some technologies preceded scientific understanding and some technologies have been used successfully without any scientific understanding. The steam engine was first used in ancient Rome, but scientific understanding of its operation in terms of the Second Law of Thermodynamics was only gained in the nineteenth century. Technologies such as the lever, the wheel and the block and tackle have been used by practical people for centuries or more, before a scientific explanation was developed. Competent electricians may know very little of the physics of electricity flow in a conductor.

An enlightened engineer of our acquaintance offers a better concept of 'technology', describing it as *'hardware, software and orgware'*. Hardware denotes a physical object, as we would expect, and 'software' is the knowledge of how to use it. 'Orgware' is something that is often overlooked by people who focus on the hardware—the organisational or institutional context that facilitates the use and dissemination of the hardware.

For example, if we consider the technologies for insulating a house, the hardware comprises the insulation materials and the software is the knowledge of which type of insulation material to use for each purpose: bulk insulation for reducing heat conduction; reflective foil for reducing radiant heat flow; and curtains with pelmets for reducing convective air flow from windows into rooms. The orgware comprises government departments, businesses and business associations, trade unions, building regulations, marketing bodies, and community and consumer groups. All three characteristics—hardware, software and orgware—are essential if a technology is to succeed.

In this section we consider technologies with a substantial component of hardware, while recognising that software and orgware must not be neglected. Thus, energy and transport systems, towns and houses, computers, communications systems and weapons are all technologies of interest in this section, but healthcare and education are not, although they do involve technologies and are relevant to other parts of this book. The classification of food and agriculture is more difficult; it receives a separate discussion in Sect. 5.1.

Technologies with low environmental impacts are sometimes described as 'green'. One of the characteristics of a green technology is that it uses raw materials that are renewable or, if that is impossible, at least abundant, as discussed in Sect. 5.2. Green technologies must be designed with long lifetimes and, if possible, for reuse. If they cannot be reused, they must be suitable for disassembly, recycling and remanufacturing. This should be standard practice, but at present, this is the exception rather than the rule. The present dominant economic system, comprising neoclassical economics and neoliberalism, is based on the ideology of leaving most decisions to the market, with the minimum number of regulations. The profits go to businesses and the adverse impacts are suffered by the people or paid for by governments. This situation can be changed, initially by community pressure on governments to enact tighter regulations and standards, and taxes on pollution, and in the longer term by replacing the destructive economic system with a constructive one, as discussed in Chap. 7.

The Sustainable Civilisation goes beyond green technologies to those that are also 'appropriate'. To unravel the latter concept, we first question the common claim that technologies are ethically neutral, that is, that they can be used either to do good or bad, for either constructive or destructive activities. While there is an element of truth in this belief, a more thoughtful view comes from the history, philosophy and sociology of science: technologies (and science itself) are selectively useful for certain activities.[68] For example, while a gun may indeed be used for shooting rabbits, its predominant use is unfortunately for shooting people. A gun that fires many bullets per second is designed entirely for shooting people.

In the USA, the ongoing tragedy of gun violence is met with the refrain by the National Rifle Association: "Guns don't kill people, people kill people." The absurdity of this slogan becomes obvious when it is placed alongside variations on the same refrain: "Burning coal doesn't emit CO_2, people emit CO_2." Whether it is mass shootings[69] or climate change, the consequences are dire where the technology is an enabler.

Claims that technologies are ethically neutral are often made by interest groups that wish to use them to control society: for example, genetic engineering of crops to control the supply of seeds to farmers[70]; information technology and artificial intelligence for monitoring and controlling people (see below). Individuals and groups in control of technologies rarely suffer the risks, which are passed on to society as a whole and the natural environment.

The uses of technologies evolve as the hardware evolves. The early huge mainframe computers were originally used to design the atomic bomb, although subsequently a much wider range of applications was found. Tiny computers, that were originally designed to fit into the warheads of guided missiles, evolved into mobile phones, which are actually very sophisticated micro-computers. Recognising that technologies are not ethically neutral, we propose that our society must design and disseminate technologies that are selectively useful for a society that is ecologically sustainable and more socially just. They must also be accessible, affordable and controllable by the people. We call them 'appropriate technologies'.

Several schools of thought contribute to this position. In his book, *Small is Beautiful*, published in 1973, E. F. Schumacher, previously Chief Economic Advisor to the UK National Coal Board, argued that the capitalist economic system depletes resources, harms the environment and devalues human life; that economic growth has adverse effects as well as positive; and that economics should serve the people instead of the reverse. This was a radical view for the time, especially for someone of his position. He posited that work must be rewarding in human terms, using human-scale, 'intermediate' technologies that are appropriate for the task.[71]

The radical technology movement was influenced by Schumacher's writing while going beyond it.[72,73] It recognises that many existing technologies do not serve our social needs well or the natural environment upon which we depend. It argues that the kind of technology adopted is determined to a large degree by the economic system, which is in turn determined to a large degree by power structures. In other words, technology choice is political. David Dickson, a former general secretary of the British Society for Social Responsibility in Science, has argued that, under capitalism,

> technological innovation was determined, not only by concern for the efficiency of production technology, but also by the requirements of a technology that maintained authoritarian forms of discipline, hierarchical regimentation and fragmentation of the labour force.[74]

To transition to an ecologically sustainable and socially just society, he argued that we need changes to the economic system, social institutions and governance as well as cleaner, human-scale technologies. The present book builds on these earlier ideas, showing that we must challenge corporate capture of the nation-state (Chap. 6) and create a new economic system (Chap. 7).

With this context, let's consider the 'appropriateness' of technology choices for several important industries.

Energy

Some technologies are so complex and dangerous that they must be tightly controlled by national governments and be subject to international agreements. In particular, nuclear power is neither accessible, nor affordable, nor controllable by the people. It requires central control by government and has to be separated from the public, because of the risk of terrorism from its materials and accidents that are devastating in health, environmental and economic terms. It serves as a means and cloak for developing nuclear weapons.[75]

In contrast, solar PV modules are much simpler, safer and more accessible—over one-quarter of Australian houses, three million in total, and many non-residential buildings already have rooftop solar electricity[76] and the number continues to grow. Community owned wind farms were the basis of a large industry in Denmark that became global (see Sect. 8.2) and community renewable electricity projects of various sizes have spread to many countries. Such projects and, more generally, community influence on government energy policy, as in the development of the *Energiewende* strategy in Germany, can be described as increasing 'energy democracy'.[77] Nowadays, environmental concern and favourable economics are driving the growth of renewable energy on small, medium and large scales, despite continuing resistance of the fossil fuel and nuclear industries and the governments and media they have captured (see Chap. 6).

Cities and their Transportation Systems

Throughout the twentieth century, many governments surrendered the planning of towns and cities to the property and motor vehicle industries. The results were sprawling developments, suburban isolation, loss of community, inadequate public transport and other infrastructure, air and water pollution, loss of biodiversity, large pockets of poverty, and ugliness. Thanks to the work of thoughtful planners such as Jane Jacobs,[78] Peter Newman and Jeff Kenworthy,[79] and Jan Gehl,[80] we have practical visions of how to renovate our towns and cities to serve people and the environment:

- taking planning out of the hands of vested interests and professional planners who put fashion before the physical and social environment
- facilitating community participation in town planning that builds diverse local communities within towns and cities

- integrating urban and transport planning
- restructuring our cities into a hierarchy of subcentres and local centres to minimise travel
- fostering public transport, cycling and walking
- concentrating housing and commercial activities around public transport hubs.

Clearly, reforming urban transportation systems must go far beyond simply replacing internal combustion engine (ICE) vehicles with EVs.

Shelter

'Housing boom!' trumpets the newspaper headlines, as if stratospheric prices are a good thing for society. Often such an article is written from the perspective of a developer or real estate speculator who profits substantially from escalating prices. It's well past time that we recognise this for what it really is: for everyone else, a 'housing boom' can be a disaster. For the majority of the world's population, shelter—home ownership and rental—is not an investment, it's a fundamental necessity, and rising prices are a sign of social disadvantage. Social justice demands that governments provide sufficient social/public housing for those in need. These should be integrated with the larger community and not relegated to the urban fringe.

The basic principles of environmentally sound housing are well understood but inadequately implemented: land subdivision to maximise orientation of houses for passive solar access; housing clustered into urban or suburban 'villages' with good access to public transport; well insulated buildings with thermal mass where appropriate; all-electric homes with rooftop solar electricity where possible; solar gardens[81] for houses and apartments unsuitable for rooftop solar PV; solar or heat pump hot water.

It is absurd and unfair that housing 'standards' in many towns specify large, minimum lot sizes and house sizes, so that purchasing a house may cost a million dollars or more. Some housing developments should provide land and infrastructure for clusters of well-designed, low-cost, tiny houses that can be wheeled into position. Government intervention in the market is needed to ensure that such sites are not forced to the urban fringe by high land prices.

Defence

The appropriate technologies must be genuinely defensive, not offensive, with military defence supplemented or replaced by social defence (see Sect. 6.4).

Information Technology

On board the Apollo 11 Lunar Module as it flew towards the Moon in 1969, were three heroes: two humans and one computer. While the Apollo guidance computer[82] is laughably small by today's standards, NASA considered it to be a crucial component. This is a striking example of how humans plus computing make an immensely powerful combination.

Since then our industrial civilisation has become considerably more complicated and would not be possible without advanced computing and related automation. A computer can process prodigious amounts of information from millions of inputs, such as weather sensors to provide accurate forecasts. That data is just one small part of the vast volumes shipped around the globe every second. Information technology (IT) has evolved into an essential tool for managing technical and financial infrastructure, and for understanding complex issues such as climate and pandemics. With the emergence of consumer electronics, it is now pervasive in every facet of our lives, from 'smart' television sets to managing healthcare.

People are gradually becoming aware of the creeping intrusion into their lives, not just of data surveillance, but also in algorithms that say what you can and cannot do, whether it's opening a car door, attending a protest rally, or online searches. The challenge is fundamentally one of control, played out between people, corporations and government. The latter theoretically represents the best interest of its population but, as discussed in Chap. 6 on state capture, governments are frequently compromised by vested interests.

Edward Snowden revealed[83] how the power of data and computing has even led governments into mass surveillance of their own citizens while they force companies to record the metadata that tracks our online activity. As security consultant Anas Baig describes it,[84] "Thanks to the new metadata retention laws passed by the Australian government, every phone call you make, text message you send, and email you write will be tracked by agencies."

While people are generally aware of these issues, there's widespread complacency as we sleepwalk into a world where everything we do is subject to scrutiny. If we are unconcerned, it's because we assume that this data is collected for a benign cause, but the cumulative effect is to equip a future totalitarian government with the ability to suppress its population. This is already happening in China where facial recognition is part of the government's "Citizen Score",[85] used to clamp down on dissent and to affect what jobs a person can or cannot have. Even in more democratic countries, companies can now use artificial intelligence bots and social media scans to rank job applicants.[86] At the George Floyd rallies of 2020, US law enforcement agencies were able to scan mobile phone records to track protestors.[87] This is a world where the past never leaves you and you could lose a career after a careless tweet.

These and other stories reveal that we are in a boiling frog situation involving the incremental loss of civil liberties. This is not a future that people want. We need community action to remind governments that their duty is to protect the rights of their citizens. Governments should not conflate 'national security' with their impulse to control while maintaining their own secrets which, as often as not, are more to do with hiding embarrassment. A glaring example is Julian Assange, who remains trapped in legal quagmire for exposing incidents such as the 'collateral murder footage'[88] from a US airstrike on 12 July 2007 in Baghdad.

The work of Assange and others shows what is possible when informed and motivated people use IT to hold governments and corporations to account. Another example is Edward Snowden's revelations that forced the Obama administration to apply 'appropriate safeguards'[89] to personal data used by intelligence services.

Europe probably provides the best approach of government to these issues. The EU General Data Protection Regulation (GDPR)[90] provides comprehensive privacy controls. Even though its jurisdiction is restricted to Europe, this and the EU's antitrust legislation have affected multinational corporations including Microsoft and IBM. Now Facebook is in their sights, causing Meta (Facebook's parent company) to issue its own threat that "*If the option to transfer, store and process data from EU users to US servers is removed, Facebook and Instagram may be shut down across Europe*".[91] Given that, with nearly three billion users, Facebook[92] is now effectively a global monopoly, there is a strong case to pursue an antitrust case (see Fig. 5.2). In fact, the US Federal Trade Commission has already been attempting for several years to split Instagram and WhatsApp from Meta.[93]

To protect privacy on a personal level, there's a vigorous community of programmers writing code to strengthen controls inside open-source web browsers such as Firefox.[95] In contrast to Google search, the DuckDuckGo[96] web search engine promises to "take back your privacy" by never tracking you.

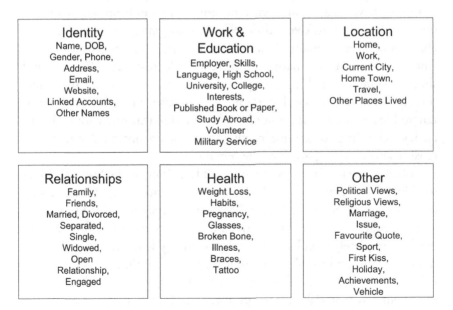

Fig. 5.2 Personal data collected by Facebook[94]

Still, for all its flaws, social media is a powerful tool for that allows activist groups to organise and strengthen the rights of citizens for any cause. The vast amount of research material online that can be discovered using search engines has become an indispensable aid, whatever your interest. Indeed, it is hard to conceive of this book being written without access to that kind of technology.

5.4   Conclusion: The Leaky Circular Economy

The ideal flow of natural resources is 'cradle to cradle', otherwise known as 'the circular economy'. That is nature's way. Nature's circular economy is possible because it is driven by the continuous input of energy from the Sun. In theory, if we humans lived truly as part of nature, perhaps we too could be entirely embedded in that circular economy. If we acted rationally, we would try to do this if we truly understood that we humans are totally dependent upon the natural world for our survival, as David Attenborough reminds us in the quotation at the beginning of the chapter.

But we have created our own economy that takes from nature's economy, while giving back very little. We are using many of nature's renewable resources at a much higher rate than nature can regenerate them. And we are using up nature's non-renewable resources. We can and should aspire to tread lightly upon our planet, substituting renewable resources for non-renewable resources wherever possible, using non-renewable resources that cannot be substituted in this way with high efficiency, according to genuine needs instead of uncontrolled wants, and reusing and recycling much of the remainder.

In theory, a circular economy is possible as long as the Sun continues to shine. In theory, with available solar and wind energy far exceeding current energy demand, all materials could be recycled. But in practice, an industrial society, even one with a much better economic system, cannot be completely circular. Although the energy losses in our leaky economic system can be replaced as long as the Sun continues to shine, there will inevitably be losses of materials at every stage of the production and use of our raw materials and the products made from them, as illustrated in Fig. 5.3. In many cases, the recovery of waste products from soils and

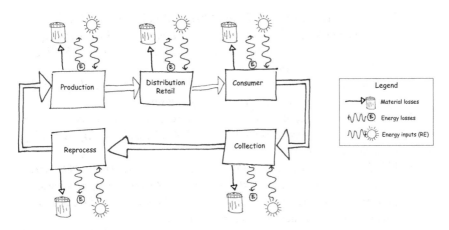

Fig. 5.3 The leaky circular economy

oceans would require exorbitant quantities of energy. A production system cannot utilise 100% of the material it receives, although this varies greatly depending on the situation. In the most efficient examples, a homogeneous material such as aluminium can be recovered and fed back into the loop. Where the material is more complex—such as a compound mix or a preassembled part—recovering unusable material becomes more difficult. More is lost when consumers discard a product rather than offering it for reuse, remanufacture or recycling. That could be simply due to inconvenience or ignorance, or, in many cases, because no alternative to discarding is available. Arguably, the most egregious barriers are deliberately placed by manufacturers that design and build products in a way that actively discourages reuse and repair.

Nevertheless, with a transformed attitude to the natural environment and transformed social and economic systems we could extend human habitation of Earth by millennia. Chaps. 6 and 7 discuss in greater depth the need to change these systems.

Notes

1. Robin McKie (2012). Interview: David Attenborough: force of nature. *The Guardian.* https://www.theguardian.com/tv-and-radio/2012/oct/26/richard-attenborough-climate-global-arctic-environment

2. Jacopo Pasotti & Elisabetta Zavoli (2018). People are living inside landfills as the world drowns in its own trash. *Huffpost*, 23 October. https://www.huffpost.com/entry/plastic-trash-pollution-landfill_n_5b9fcc13e4b013b0977d47ce?guccounter=1.

3. William McDonough & Michael Braungart (2002). *Cradle to Cradle: Remaking the Way We Make Things*. North Point Press.

4. Jouni Korhonen et al. (2018). Circular economy: the concept and its limitations. *Ecological Economics* 143:37–46. https://doi.org/10.1016/j.ecolecon.2017.06.041.

5. US United States Environmental Protection Agency. *National Overview: Facts and Figures on Materials, Wastes and Recycling*. https://www.epa.gov/facts-and-figures-about-materials-waste-and-recycling/national-overview-facts-and-figures-materials.

6. Korhonen et al. *op. cit.* Fig. 2.

7. Sam Davis (2022). The Keep America Beautiful campaign and greenwashing. https://www.dogwoodalliance.org/2022/06/the-keep-america-beautiful-campaign-and-greenwashing/; Finis Dunaway (2017). The 'Crying Indian' ad that fooled the environmental movement. *Chicago Tribune* 21 November. https://www.chicagotribune.com/opinion/commentary/ct-perspec-indian-crying-environment-ads-pollution-1123-20171113-story.html.

8. *The Observer* (2022). https://sacobserver.com/2022/04/world-trade-head-predicts-food-riots-in-poor-countries-due-to-ukraine-war.

9. CNBC (2022). www.cnbc.com/2022/04/28/russia-ukraine-war-threatens-the-middle-easts-food-security.html.

10. Julian Cribb (2019). *Food or War*. Cambridge University Press.

11. IPCC (2019). Special Report—Climate Change and Land. https://www.ipcc.ch/srccl/.

12. Walter Willett et al. (2019). Food in the Anthropocene: the EAT-Lancet Commission on healthy diets from sustainable food systems. *Lancet* 393:447–492. https://doi.org/10.1016/S0140-6736(18)31788-4.

13. Willett et al. *op. cit.*

14. Society for the Protection of Underground Networks (SPUN). https://www.spun.earth/networks/mycorrhizal-fungi.

15. Peter Aldhous (2014). We are killing species at 1000 times the natural rate. *New Scientist*, 29 May. https://www.newscientist.com/article/dn25645-we-are-killing-species-at-1000-times-the-natural-rate.

16. Jonathan Sanderman, Tomislav Hengl & Gregory Fiske (2017). Soil carbon debt of 12,000 years of human land use. *Proceedings of the National Academy of Sciences* 114 (36): 9575–9580, 2017. https://doi.org/10.1073/pnas.17061031.

17. Jerry Melillo & Elizabeth Gribkoff (2021). MIT Climate Portal. https://climate.mit.edu/explainers/soil-based-carbon-sequestration.

18. Gabe Brown (2018). *Dirt to Soil: One family's journey into regenerative agriculture.* Chelsea Green; David Montgomery 2017). *Growing a Revolution: Bringing our soil back to life.* New York & London: WW Norton; Mark Shepard (2013). *Restoration Agriculture: Real-world permaculture for farmers.* Austin, Texas: Acres USA.

19. NRDC (2021). https://www.nrdc.org/stories/regenerative-agriculture-101.

20. Deakin University (2015). Media release. https://www.deakin.edu.au/about-deakin/news-and-media-releases/articles/deakin-scientists-look-at-role-of-wetlands-in-battle-against-climate-change.

21. Sarah Derouin (2017). https://eos.org/articles/study-finds-that-coastal-wetlands-excel-at-storing-carbon.

22. Mark Mulligan & Sophia Burke (2009). Global database of dams. www.ambiotek/tropicalhydrology.

23. International Rivers website. Environmental impacts of dams. https://archive.internationalrivers.org/environmental-impacts-of-dams.

24. Edson Krenak Naknanuk (2021). The Amazon is dirty, our rivers and fish are contaminated, everyone is sick. https://www.culturalsurvival.org/news/amazon-dirty-our-rivers-and-fish-are-contaminated-everyone-sick; Fabio Zuker (2021). Amazon tribe suffers mercury contamination as illegal mining spreads. *Reuters,* 17 December. https://www.reuters.com/article/us-climate-brazil-amazon-mercury-idUSKBN2IW03Q.

25. Rivers of Carbon. https://riversofcarbon.org.au/resources/the-science-of-riparian-carbon.

26. The Fish Site (2009). Interview with Grimur Validmarsson. https://thefishsite.com/articles/fish-in-the-global-food-chain-challenges-and-opportunities.

27. Australian Broadcasting Corporation (2021). The Science Show. Interview with Daniel Pauly. 13 February. https://www.abc.net.au/radionational/programs/scienceshow/90-of-all-big-fish-have-been-taken-from-the-oceans/13149932.

28. Susan Lawler (2012). *The Conversation*. https://theconversation.com/super-trawlers-the-juggernauts-of-the-oceans-environmental-economic-and-political-devastation-8812.

29. Christopher Golden et al. (2016). Nutrition: Fall in fish catch threatens human health. *Nature* 534:317–320. https://doi.org/10.1038/534317a.

30. United Nations Global Compact. Sustainable seafood. https://unglobalcompact.org/take-action/ocean/communication/sustainable-seafood.

31. Lauren Sara McKee (2020). *Massive Science*. https://massivesci.com/articles/aquaculture-fish-antimicrobial-resistance-antibiotics-sustainability-ocean.

32. Salmon Facts (2016). https://salmonfacts.com/what-eats-salmon/is-salmon-feed-sustainable.

33. Ellen Macarthur Foundation. https://ellenmacarthurfoundation.org/circular-examples/regenerative-ocean-farming.

34. WWF. https://wwf.panda.org/discover/our_focus/forests_practice/importance_forests.

35. WWF. https://explore.panda.org/forests.

36. Luke Taylor (2022). The Amazon rainforest has already reached a crucial tipping point. *New Scientist* https://www.newscientist.com/article/2336521-the-amazon-rainforest-has-already-reached-a-crucial-tipping-point.

37. Rachel Graham (2022). Amazon deforestation: why is the rainforest being destroyed?. https://sentientmedia.org/amazon-deforestation.

38. Luciana Gatti et al. (2021). Amazonia as a carbon source linked to deforestation and climate change. *Nature* 595:388–393. https://doi.org/10.1038/s41586-021-03629-6.

39. WWF. https://www.wwf.org.uk/press-release/illegal-deforestation-report-brazil; Francisco Filho et al. (2021). https://www.climatepolicy-initiative.org/publication/the-economics-of-cattle-ranching-in-the-amazon-land-grabbing-or-pushing-the-agricultural-frontier.

40. Al Jazeera (2022). News. https://www.aljazeera.com/news/2022/7/20/brazil-authorities-doing-little-to-prevent-deforestation-report.

41. Larissa Basso & Cristina Inoue (2021). *The Conversation*. https://theconversation.com/even-if-bolsonaro-leaves-power-deforestation-in-brazil-will-be-hard-to-stop-163964.

42. NASA Earth Observatory, https://earthobservatory.nasa.gov/images/147946/fires-raged-in-the-amazon-again-in-2020.

43. Divina Ramirez (2020). *Climate Science News.* https://www.climate-sciencenews.com/2020-08-21-beetle-infestations-are-killing-forests.html.
44. Despite the name, the fires actually commenced in mid-Winter 2019.
45. Mark Binskin et al. (2020). *Royal Commission into National Natural Disaster Arrangements—Report.* https://naturaldisaster.royalcommission.gov.au/publications/royal-commission-national-natural-disaster-arrangements-report.
46. Australian Government, Department of Climate Change, Energy, the Environment and Water (2020). https://www.dcceew.gov.au/climate-change/publications/estimating-greenhouse-gas-emissions-from-bushfires-in-australias-temperate-forests-focus-on-2019-20.
47. Australian Government, Department of Climate Change, Energy, the Environment and Water (2022). https://www.dcceew.gov.au/climate-change/publications/national-greenhouse-gas-inventory-quarterly-update-december-2021.
48. FAO. 2022. *In Brief to The State of the World's Forests 2022. Forest pathways for green recovery and building inclusive, resilient and sustainable economies.* Rome, FAO. https://doi.org/10.4060/cb9363en.
49. Forest Stewardship Council, https://fsc.org/en/forest-management-certification.
50. Timothy Beatley, Heath Boyer & Peter Newman (2017). *Resilient Cities: Overcoming fossil-fuel dependence.* 2nd ed., Island Press.
51. World Resources Institute. https://www.wri.org/insights/progress-must-speed-protect-and-restore-forests-2030.
52. Adam Sherwin (2020). https://inews.co.uk/culture/television/sky-cobra-disaster-drama-chernobyl-solar-storm-382670.
53. ArchDaily. https://www.archdaily.com/893442/cross-laminated-timber-clt-what-it-is-and-how-to-use-it; TPM Builders, https://www.tpmbuilders.com.au/cross-laminated-timber-buildings-engineered-wood.
54. Will Steffen et al. (2016). How defining planetary boundaries can transform our approach to growth. *Solutions* 22:2 . https://thesolutionsjournal.com/2016/02/22/how-defining-planetary-boundaries-can-transform-our-approach-to-growth.
55. National Institutes of Health. https://ods.od.nih.gov/factsheets/Phosphorus-HealthProfessional.
56. Department of Primary Industries. https://www.dpi.nsw.gov.au/agriculture/soils/more-information/improvement/phosphorous.

57. Sean Comber et al. (2013). Domestic source of phosphorus to sewage treatment works. *Environmental Technology* 34:1349–1358. https://doi.org/10.1080/09593330.2012.747003.

58. Comber et al. *op. cit.*

59. A. Edward Johnston et al. (2014). Chapter Five—Phosphorus: its efficient use in agriculture. *Advances in Agronomy* 123:177–228. https://doi.org/10.1016/B978-0-12-420225-2.00005-4.

60. American Society of Agronomy (2019). Reduce, reuse, recycle: the future of phosphorus in agriculture. *SciTechDaily,* https://scitechdaily.com/reduce-reuse-recycle-the-future-of-phosphorus-in-agriculture

61. Datu Buyung Agusdinata et al. (2018). Socio-environmental impacts of lithium mineral extraction: towards a research agenda. *Environmental Research Letters* 13:123001. https://iopscience.iop.org/article/10.1088/1748-9326/aae9b1/pdf.

62. Peter Greim et al. (2020). Assessment if lithium criticality in the global energy transition and addressing policy gaps in transportation. *Nature Communications* 11:4570. https://doi.org/10.1038/s41467-020-18402-y.

63. Thomas Vranken (2020). Critical raw materials in Li-ion batteries. *InnoEnergy,* https://www.innoenergy.com/media/5817/critical-raw-materials-in-li-ion-batteries.pdf.

64. Sophie Vorrath (2021). https://reneweconomy.com.au/battery-recycling-plant-starts-shredding-in-germany-using-australian-technology.

65. Ana Lejtman (2022). Climatebiz. https://climatebiz.com/lithium-battery-alternatives.

66. Tobias Junne et al. (2020). Critical materials in global low-carbon energy scenarios: The case for neodymium, dysprosium, lithium, and cobalt. *Energy* 211:118532. https://doi.org/10.1016/j.energy.2020.118532.

67. *Encyclopedia Britannica*, 7 Apr. 2022, https://www.britannica.com/technology/technology.

68. Brian Martin (1983). The selective usefulness of science. *Queen's Quarterly* 90(2):489–496. https://documents.uow.edu.au/~bmartin/pubs/83qq.html.

69. An even more malignant example is the use of drone weapons where the operator might not be even situated in the same continent. And now automated, AI killing machines are on the immediate horizon, making attempts to ban them by international treaty urgent.

70. Organic Consumers Association. *Seeds of Evil: Monsanto and genetic engineering.* https://www.organicconsumers.org/news/seeds-evil-monsanto-and-genetic-engineering.

71. Ernst Schumacher (1973). *Small is Beautiful: A study of economics as if people mattered.* Harper Collins.
72. Godfrey Boyle, David Elliott & Robin Roy (1977). *The Politics of Technology.* Longmans & The Open University Press.
73. Peter Harper, Godfrey Boyle and the editors of *Undercurrents* (eds) (1976). *Radical Technology: Food and Shelter, Tools and Materials, Energy and Communications, Autonomy and Community.* Penguin.
74. David Dickson (1974). *Alternative Technology and the Politics of Technical Change.* Fontana, Chapter 3.
75. See The Nuclear Weapon Archive https://nuclearweaponarchive.org/; Institute for Science and National Security https://isis-online.org/;; and articles by David Albright.
76. Australian Government, Department of Climate Change, Energy, the Environment and Water (2022). https://www.energy.gov.au/news-media/news/australia-achieves-3-million-rooftop-solar-pv-installations.
77. Craig Morris & Arne Jungjohann ((2016). *Energy Democracy: Germany's Energiewende to renewables.* Palgrave Macmillan, especially Chapter 6 & 10.
78. Jane Jacobs (1961). *The Death and Life of Great American Cities.* Random House.
79. Peter Newman & Jeffrey Kenworthy (1999). *Sustainability and Cities: Overcoming Automobile Dependence.* Island Press; Peter Newman & Jeffrey Kenworthy (2015). *The End of Automobile Dependence.* Island Press.
80. Jan Gehl & Lars Gemzøe (2004). *Public Spaces, Public Life.* Danish Architectural Press.
81. Solar garden. https://haystacks.solargarden.org.au.
82. Curious readers will enjoy this description of the ACM: David Szondy (2019). Apollo's brain: The computer that guided man to the Moon. *New Atlas,* 9 June. https://newatlas.com/apollo-11-guidance-computer/59766.
83. Glenn Greenwald (2014). *No Place to Hide: Edward Snowden, the NSA, and the U.S. Surveillance State.* Henry Holt and Company.
84. Anas Baig (n.d.). Key Details on the Australian Meta-Data Retention Law. *Infosecurity,* http://www.infosecurity-magazine.com/blogs/metadata-retention-law.
85. CBS News (2018). https://www.cbsnews.com/newyork/news/china-assigns-every-citizen-a-social-credit-score-to-identify-who-is-and-isnt-trustworthy.

86. Stephen Jones (2021). *Business Insider*, 8 September. https://www.businessinsider.com/ai-recruitment-tools-cv-scanners-automated-hiring-overlook-hidden-workers-2021-9?op=1.

87. Mike Fong (2020). *Forbes*, 21 August. https://www.forbes.com/sites/forbestechcouncil/2020/08/21/going-to-a-protest-keep-your-smart phone-from-being-used-as-a-tracking-device/?sh=73c8e55c6fb2.

88. Dan Froomkin (2010). WikiLeaks VIDEO Exposes 2007 'Collateral Murder' In Iraq. *Huffpost*, 5 June, https://www.huffpost.com/entry/wikileaks-exposes-video-o_n_525569

89. Sarah Childress (2015). PBS, 9 February. https://www.pbs.org/wgbh/frontline/article/how-the-nsa-spying-programs-have-changed-since-snowden.

90. Privacy Shield Framework (n.d.). https://www.privacyshield.gov/article?id=European-Union-Data-Privatization-and-Protection.

91. Dashveenjit Kaur (2022). *TechHQ*, 7 February. https://techhq.com/2022/02/europes-data-laws-might-see-meta-shutting-down-facebook-instagram-there.

92. S. Dixon (2022). *Statista*, 28 July. https://www.statista.com/statistics/264810/number-of-monthly-active-facebook-users-worldwide.

93. Harry Pettit (2022). *The U.S. Sun*, 12 January. https://www.the-sun.com/tech/4446512/facebook-lose-instagram-whatsapp-lawsuit-apps-change.

94. Derived from Andreea M Belu (2017). The massive data collection by Facebook—Visualised. *DataEthics.eu*, 26 June. https://dataethics.eu/facebooks-data-collection-sharelab

95. A typical example is the "Facebook Container" (see https://addons.mozilla.org/en-US/firefox/addon/facebook-container) that prevents Facebook from tracking you around the web. For a small fee, you can use a "VPN" to encrypt your internet traffic and disguise your online identity.

96. https://duckduckgo.com.

Reference

Weblinks accessed 27/10/2022.

6

Cutting the Bonds of State Capture

A system is corrupt when it is strictly profit-driven, not driven to serve the best interests of its people. (Suzy Kassem[1])

President Franklin D. Roosevelt is supposed to have told a delegation: "*OK, you've convinced me. Now get out there and make me do it!*". A community that wants to get a public interest policy accepted by governments, against opposition from vested interests, must demand, campaign and struggle for it with great determination. Sometimes these efforts bring success, such as the abolition of government support for slavery, votes for women in many countries, civil rights for black Americans, and the non-violent overthrow of dictatorships in the Philippines and Argentina.

For many other issues, even in countries with a long, proud tradition of democracy, governments sometimes implement policies opposed by the majority of their citizens. An example is the British government's persistence with the Trident nuclear strike force despite majority opposition manifest in many public opinion polls.[2] Consider further that the United Kingdom, through its membership of NATO, implicitly supports the first use of nuclear weapons, although a survey found that two-thirds of the British public oppose first use.[3] Numerous polls found that Australians did not wish to participate in the US-led attack on Iraq,[4] but the Prime Minister decided otherwise. A poll conducted in 2021 found

© The Author(s), under exclusive license to Springer Nature Singapore Pte Ltd. 2023   **125**
M. Diesendorf, R. Taylor, *The Path to a Sustainable Civilisation*,
https://doi.org/10.1007/978-981-99-0663-5_6

that two-thirds of Australians favour increased action on climate change and the adoption of renewable energy,[5] but the then Coalition government refused to increase its tiny greenhouse gas target for 2030 and persisted in its plan for a 'gas-led recovery' from the pandemic.[6] That government lost office at the May 2022 election and so democracy eventually worked to a limited degree. However, the new Labor government legislated a stronger target that still falls far short of the requirement from climate science and persists in supporting the development of new coal mines and gas fields.[7]

How can we explain these and other decisions by governments that give lip service to democratic decision-making while defying it in practice? How can we explain the resistance by governments to the policies recommended in Chaps. 4 and 5 to address specific environmental and social justice problems, some of which are existential for human civilisation? These policies are rational and in the public interest and most have been advocated by community organisations for many years.

This chapter identifies the fundamental driving forces responsible for all these problems (Sect. 6.1) and recommends over-arching strategies and policies (defined in Box 6.1) that could help solve them simultaneously (Sects. 6.2 and 6.3). By 'over-arching', we mean that these policies would be beneficial for many issues, not just a single issue. Because each of these policies has multiple benefits, citizen pressure to implement them can be applied by many more organisations and individuals than exist for single issues. Activist groups concerned with specific issues can cooperate on over-arching strategies and policies to strengthen democracy, change the economic system and transition to a sustainable civilisation.

Box 6.1 What Is a Policy?

A policy is a statement of intention. To be effective, it must be coupled with a strategy and process for implementation, monitoring and enforcement. To be truly democratic, a government policy must be designed with substantial public input.

For socioeconomic and political changes within a country, citizen pressure can be applied to national, state/provincial and local spheres of government, as well as to business, industry, and professional and other organisations. To address global problems such as climate change, biodiversity loss, deforestation and de facto slavery, citizens can apply pressure to national governments for new international agreements, as well as for strengthened existing agreements at global scale (Sect. 6.4). Although it's very difficult to gain strong international agreements on sustainability, their need must be recognised and negotiations must commence as soon as possible. Where governments fail to act, direct citizen action can create new projects and new institutions independent of governments (Chap. 8).

6.1   The Roots of the Crises

Many community-based non-government organisations (NGOs) have been working for decades to remove or reduce the many threats to ecological sustainability, social justice, human health and world peace. They campaign to protect Earth's climate, biodiversity, forests, atmosphere, soils and waterways; and to control harmful technologies, gambling, cruelty to animals, and pollution by chemicals, noise and ionising radiation. They strive for equal human rights for women and minorities, and nuclear disarmament and peace between nations. They resist dictatorial governments and other powerful interest groups. They campaign to stop slavery and other forms of economic exploitation, and work towards equal opportunity for all people to have a high quality of life.

Without NGOs and their specific campaigns, these threats and injustices would be much worse than at present. When we, the authors, feel depressed about the state of the world and humanity, we visit the website of the Right Livelihood Award,[8] also known as the Alternative Nobel Prize, and are inspired by the courageous and visionary achievements of individuals and (initially) small groups in striving for a more just, peaceful and sustainable world. Despite these efforts, several issues—such as climate change, biodiversity loss and the threat of nuclear war—have become existential threats. Therefore, campaigns on specific issues must be supplemented by over-arching collective campaigns against the principal driving forces behind all or most of these issues.

The fundamental forces can be described as the desire for power over others and/or the endless accumulation of wealth. These drives in the psychology of some people are not amenable to change on a social or political scale.[9] They lie beneath the surface, as sketched in Fig. 6.1. The desire for power over others, that is inadequately limited by social institutions, has led to patriarchal entitlement and the subjugation of women including systemic violence against women globally. It has produced 'strong man' autocrats and dictators who desire to control what 'the people' think and do, and who brook no dissent. The desire for both power over others and great wealth have led to the exploitation of the planet and people on a global scale.

Human society can guide and control the expression, the manifestations, of those fundamental destructive forces that are above the surface. At the first level above the surface are capitalism, autocracy/oligarchy and patriarchy (defined in Box 6.2).

Box 6.2 Definitions of Capitalism, Autocracy, Oligarchy and Patriarchy

Capitalism is an economic and political system in which a country's trade and industry are controlled by private owners for profit, rather than by the state or workers. This definition recognises that capitalism is not just an economic system but is also politically determined. The private owners of the means of producing and distributing goods (factories, technologies, land, transportation system) are a small minority of people. They employ the majority of people, who sell their labour in return for a salary or wage. The private owners live off part of the profits and invest the rest for further accumulation of wealth. It is widely believed that capitalism depends on endless economic growth, although it's unclear whether this has been proven. Capitalism can thrive under representative government or autocratic rule. In state capitalism, for example, in the People's Republic of China, an autocratic state exercises strong control over the economy and the people.[10]

Autocracy is defined as *a system of government in which a single person or party (the autocrat) possesses supreme and absolute power.*

Another, related expression of greed and the lust for power is an *oligarchy*, defined as a *small group of people having control of a country or organisation.*

Patriarchy is the *systematic domination of women by men in some or all of society's spheres andinstitutions:* for example, *the family, government, institutions and employment.*

Drawing upon the quotation by Suzy Kassem at the beginning of the chapter, one could add that each of these institutions is often associated with corruption.

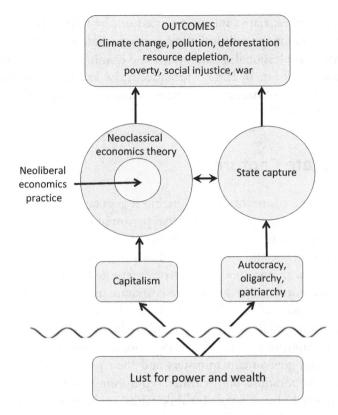

Fig. 6.1 Driving forces of environmental destruction and social injustice

Capitalism and autocracy/oligarchy have been resisted for centuries and so are unlikely to disappear in time to deal with the existential threats faced by human civilisation. While such resistance can and should continue, a more promising point of attack is at the next level up in Fig. 6.1, which has two key elements, neoliberal economics practice (also known as neoliberalism) and state capture by vested interests.

This chapter examines state capture and how to reduce it. In the context of this book, 'the state' means *the government of a country together with the public service, the military and police that implement and protect its policies.* State capture is *the exercise of power by private actors—through control over resources, threat of violence, or other forms of influence—to shape government policies or implementation in service of their narrow interest.* In various countries, state capture is carried out by large corporations, the military, religious and political movements, foreign governments, powerful families, professional organisations and criminal networks. We focus here on state capture by large corporations, also known as corporate capture. The failure of neoliberal economics and better alternatives are discussed in Chap. 7.

6.2   State Capture

State capture is a form of corruption, but it goes far beyond petty corruption where, for example, an individual person or business pays a government minister or public official to influence a government decision. State capture is systemic and well-organised by people who have an established relationship with one another. It involves repeated transactions, often on an increasing scale, and can involve corporate management of all major political parties, public officials, mass media, social media, police and institutions originally designed to regulate the industry concerned. State capture is implemented by political donations, lobbying, 'revolving door' jobs between government ministers and their political advisers on one hand and corporations on the other, repurposing institutions, astroturfing,[11] 'research' by self-styled 'think tanks', policymaking, and public influence campaigns.[12,13,14]

Large multinational organisations dominate globally in the mining, armaments, forestry, agriculture, pharmaceuticals, property, banking/finance and information technology industries. They are the most prominent in terms of wealth and influence in politics and society. Economists and professional people trained in neoliberal thinking exercise strong influence over the policies of many governments on behalf of large corporations. These corporations and professionals are responsible for most state capture. For this reason, state capture is often called 'corporate capture of the state' although other organisations and foreign governments also conduct state capture. One outcome of state capture is the huge level of subsidies to the fossil fuel industries discussed in Sect. 4.3.

This chapter puts forward a strategy for collectively challenging and weakening state capture, thereby strengthening democracy within countries. It also proposes an international strategy. First, we present two case studies of state capture in Australia: capture by the fossil fuel industries and by the armaments industry. In each case, the outcomes are the same, that is, the industry determines government decisions, supported to a large degree by the opposition party, although the respective industry portfolios of capture tactics differ to some degree between the two case studies.

Australia's decisions in these sectors matter globally. Although only a middle power, Australia is one of the world's biggest exporters of coal and natural gas. Therefore, its climate and energy policies can substantially influence global greenhouse gas (GHG) emissions. In terms of defence (or, more accurately, war-making), Australia is playing the role of the USA's 'deputy sheriff'[15] in the Asia-Pacific region; its weapons systems are mostly those of the USA and the UK.

Australia is a parliamentary democracy in the British tradition. Both at the federal/national level and in the states, it has two major political parties, the conservative Liberal-National Coalition and the less conservative Australian Labor Party (ALP). It has two Houses of Parliament, the House of Representatives and the Senate. The voting system[16] for the House of Representatives ensures that one or other of the two major parties is always in government. Among several smaller parties, the largest is The Greens. The smaller parties and Independents can sometimes join with the Opposition to reject legislation, especially in the Senate.

The Centre for Public Integrity estimates that, at the federal level, both major political parties received donations totalling more than A\$4.2 billion over two decades, of which about A\$1.5 billion could not be traced to a source.[17] Only individual donations above A\$14,500 must be declared and these are only reported annually. Multiple donations below the threshold can be made to state and territory branches of the same party without a requirement to declare the total. Substantial donations are also given indirectly via other organisations, such as industry associations, trade unions (to the ALP), investment funds, and political party fund-raising 'forums' which charge high membership fees for access to decision-makers.

Case Study: Australian State Capture by the Fossil Fuel Industries

In the early 2000s, a group representing big emitters of greenhouse gases, called the Australian Industry Greenhouse Network, exercised control over the Coalition Government's climate and energy policies through all the methods listed above. They were so successful that they referred to themselves internally as the 'greenhouse mafia'.[18,19] But they had not yet captured the Opposition. The election of the ALP government in 2007 resulted in Australia signing the Kyoto Protocol and legislating several policies to assist the growth of renewable energy. But when Prime Minister Kevin Rudd attempted to introduce a mining tax in 2010, the fossil fuel industry ran a A$22 million campaign that helped to remove him from the top job.[20] He was replaced in the ALP government by Julia Gillard, whose government did not pursue an effective mining tax.

However, under pressure from the Greens, who held the balance of power, the ALP government reluctantly introduced a carbon price in 2012. It also set up the Australian Renewable Energy Agency (ARENA), to award grants for research, development and demonstration, and the Clean Energy Finance Corporation (CEFC), to make financial investments in renewable energy and energy efficiency projects. The fossil fuel industry responded by putting substantial resources into the 2013 election campaign, which was supported by the Murdoch press (News Corp), and the ALP lost the election. The new Coalition government immediately removed the carbon tax before it could evolve into a tradable emissions scheme as the ALP had intended. Subsequently, the ALP, which returned to government in May 2022, has been cowed and has only announced climate and energy policies that are slightly better than those of the Coalition.

After leaving Parliament, Ministers from both major parties receive generously paid 'revolving door' jobs in industries for which they had been responsible. These include permanent senior appointments, consultancies, lobbyist roles and media commentators. For example, Ian Macfarlane, immediately after his resignation as Industry Minister of the Coalition government in 2016, was hired as the CEO of the Queensland Resources Council, the principal fossil fuel lobby organisation in that state, and he occupies a board position with Woodside, a multinational fossil fuel business focusing on gas. Martin Ferguson, who served as Minister for Resources and Energy in the ALP government from 2007 to 2013, almost immediately took up the role of chairman of the Advisory Council of the Australian Petroleum Production & Exploration Association (APPEA), and took up a non-executive directorship of Seven Group Holdings, which is owner of Seven West media and major shareholder in Beach Energy, an oil and gas company. Revolving doors rotate in both directions. The Chief of Staff of the Coalition Prime Minister, Scott Morrison (August 2018–May 2022), was John Kunkel, former deputy CEO of the Minerals Council of Australia (MCA), one of the principal lobby and public campaign organisations for fossil fuels. A senior adviser to the Prime Minister was Brendan Pearson, a former CEO of the MCA.[21]

The fossil fuel industries have pressured the government to repurpose key institutions. In 2012, the Institute of Public Affairs,[22] a right wing 'think tank' campaigning for inter alia fossil fuel and other mining industry positions, published a list of policy demands[23] for the Liberal Party, which was then in opposition. When the Liberal-National Coalition was elected to government in 2013, it implemented many of these policy demands, including regulations that try to force ARENA to fund carbon capture and storage projects, and an unsuccessful attempt to force CEFC to invest in fossil gas and nuclear power projects.[24] To guide an economic recovery from the COVID-19 pandemic, the Coalition Government set up a Commission dominated by the gas industry which unsurprisingly recommended a 'gas-led recovery', that is, that new gas basins be opened by public funding for pipelines, but did not recommend support for renewable energy.[25]

Industry has also been pressuring the government to repurpose the legal system to criminalise nonviolent protest and to legislate against 'secondary boycotts', that is, attempts to influence the actions of one business by exerting pressure on another business. For example, community NGOs were practising secondary boycotts by demanding that companies do not provide insurance or banking for projects such Adani's huge Carmichael coal mine. The then Prime Minister announced that the government would try to find a way to compel businesses legally to provide services to fossil fuel companies.[26]

Studies of the coverage of climate change by Australian newspapers by the Australian Centre for Independent Journalism[27] and by community advocacy group GetUp[28] found a large proportion of articles and editorials dismissed or questioned whether human activity was causing climate change. The principal media bias came from News Corp, which owns the majority of Australian newspapers. These newspapers also publish frequent attacks on renewable energy.

In Australia, state capture by the fossil fuel industry was almost complete at the federal government level at the end of 2021. It remains to be seen whether the new ALP government, elected in May 2022, can throw off some of the industry's bonds.

The above examples are just a small sample of the evidence—for more details, see the reports cited above[29] and Lucas (2021).[30]

Case Study: Australian State Capture by the Armaments Industry and a Foreign Government

Since the armaments industry sells its products directly to the Australian Government, it doesn't have to build a positive public image. Its tactics are concentrated on lobbying the government and organisations that advise the government, revolving door jobs, biased research and policy-making, and briefing commentators who promote the alleged need for new weapons for national security.

The industry has benefitted from the notion, believed by successive governments and the majority of the public, that Australia must rely on the protection of the USA. This 'justifies' the Australian government in purchasing expensive military hardware, much from the USA, and participating in wars led by the USA in distant lands. Australia is the second largest customer of the US arms industry, according to research by the Stockholm International Peace Research Institute.[31]

A key source of confidential advice to the Australian Government on defence and military matters is the Australian Strategic Policy Institute (ASPI). While it claims to be independent, ASPI is funded by the US and other governments, and the armaments industry, as well as the Australian government.[32]

Thus, Australia has been captured by the military-industrial complex of the USA. Further evidence of this capture is the announcement of a new strategic partnership between Australia, the United Kingdom and the United States—the AUKUS Agreement.[33] A key part of this partnership has been the recent abrupt cancellation of Australia's contract to buy 12 diesel-powered submarines from France and the announcement to replace them with nuclear-powered submarines purchased from the USA or UK. This reflects the change from the type of submarine most suitable for defending the coastline to one that is most suitable for participating in operations with the USA in a distant region, such as the South China Sea. Furthermore, the use of highly enriched uranium as the fuel in the US and UK nuclear-powered submarines creates the risk of future nuclear weapons proliferation by Australia and the possibility of triggering a regional nuclear arms race. Until now, the only countries with nuclear-powered submarines have been nuclear weapons powers.[34]

The switch from French to US/UK nuclear-powered submarines can be understood in the light of the recent revelation that six retired US admirals have worked for the Australian government since 2015, including one who was employed for two years as the Deputy-Secretary of Defence. "In addition, a former U.S. secretary of the Navy has been a paid adviser to three successive Australian prime ministers." This was revealed by investigative journalism by the *Washington Post*.[35] As a former Australian Defence official pointed out:

> The Washington Post has disclosed the extent to which Australia's sovereignty and security has been handed over to the Pentagon. On the evidence it appears that the nuclear powered submarine decision process was heavily influenced by a clique of former US Navy Admirals with potential conflicts of interest, and who were generously paid by the Australian government. What confidence can Australians have in the soundness of this opaque, over-priced, strategically unjustifiable, and massively underspecified project?[36]

Meanwhile, the US government is funding the expansion of an air force base in Australia's Northern Territory to allow it to house six B-52 nuclear weapons-capable bombers. Both this decision and the nuclear-powered submarine decision were made without any debate in Parliament.

Revolving doors are an important feature of state capture by the weapons industry. According to Michael West Media:

> In June 2018, Mark Binskin was Chief of the Defence Force when BAE Systems Australia was awarded the $35 billion Future Frigate contract, the largest surface warship program in Australia's history. The following month Binskin retired. He has since been appointed in a non-executive director role with BAE Systems. The contract for the $1.2 billion upgrade of the Jindalee Operational Radar Network was also awarded to BAE in the final months of Binskin's tenure.[37]

Former head of the Australian Security Intelligence Organisation (ASIO), Duncan Lewis, joined the Australian board of Thales, a French arms and security multinational and a top three Australian defence contractor, five months after leaving ASIO.[38] A possible outcome of revolving door appointments and lobbying was the unexpected award of a A$1.3 billion contract to Thales for medium-weight armoured vehicles. The procurement was so unexpected that it was the subject of an investigation by the Australian National Audit Office.[39]

In other revolving door appointments, Brendan Nelson, former Liberal party leader, Defence Minister and Director of the Australian War Memorial, was appointed president of Boeing Australia, New Zealand and South Pacific, a top five contractor to Defence.[40] The former Coalition Minister for Defence, Christopher Pyne, accepted a job with the large consulting corporation EY to help grow its defence business. The appointment was announced a little after a month after Pyne retired from federal Parliament.[41] Some revolving doors with possible conflicts of interest are illustrated in Fig. 6.2.

While being concerned about potential conflicts of interests in revolving door appointments, we do not question the motives of the people involved.

More evidence of state capture of Australia by the armaments industry has been published in the reports cited above by Michael West Media and the Australian Democracy Network.

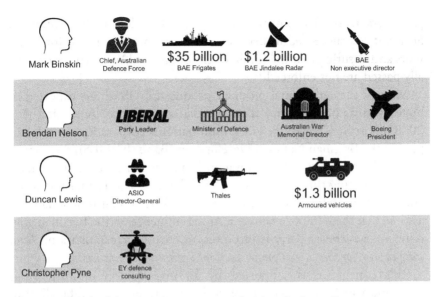

Fig. 6.2 The revolving door in armaments decisions in Australia

Next, we discuss state capture of the Global South by large corporations that are sometimes backed by foreign governments.

6.3   State Capture of the Global South

The drivers of destruction of the rainforest of Indonesia are the international demands for palm oil and rubber together with the domestic demands for rice and cassava. In the African rainforest, the main driver is mining. In the Brazilian rainforest, the international drivers are beef, minerals, timber and soybeans (see Box 5.3). Here we consider Brazil as a case study of state capture, although similar accounts, with the same conclusion, can be given for other countries.

The continuing clearing and burning of the Brazilian rainforest, despite its severe impacts on the national and global environment and the Indigenous inhabitants, can be understood as the result of state capture by agribusiness interests. The Ruralist Caucus is a group of members of Brazil's parliament from several political parties representing big rural producers. In 2016 it comprised 207 out of 513 federal parliamentarians. So far, it has succeeded in delaying environmental legislation for many years and ensuring that the (limited) environmental laws that are eventually passed are not enforced.[42]

Most of the agricultural products produced by Brazil are exported to huge overseas national and transnational corporations. According to a 2008 report by Greenpeace USA, multinational corporations, funded in part by the Brazilian National Development Bank (BNDES), the finance arm of the Ministry of Development, Industry and Foreign Trade, have been supporting the expansion of the global beef and leather industries and hence the deforestation of Amazonia.[43] Therefore, it is reasonable to assume that these importers have a strong influence on the Brazilian government, whichever party is in power, to encourage exploitation. *Thus, state capture of the Global South overlaps with neo-colonialism.*

This conclusion is supported by an investigation into the Atlas Network, which has reshaped political power to favour neoliberalism in many countries of the Global South, including Brazil. Atlas receives funding from foundations set up by the billionaire Koch brothers and, it is alleged, from the U.S. State Department and the National Endowment for Democracy, a critical arm of American soft power. Together with its affiliated charitable foundations, Atlas has given hundreds of grants to free-market 'think tanks' in Latin America.[44]

Part of the solution may come from international investors—those concerned about the financial and reputational risks inherent in environmental and social impacts. In Brazil, most progress has been made in the securities market and banking sector where international investors have encouraged some companies to develop and disclose voluntary socioeconomic policies based on international accepted standards.[45] Internationally, the campaigns for environmental and social certification of products and for divestment from corporations that don't comply, are beginning to have an effect.

6.4   Strategy to Weaken State Capture

Case studies of state capture by the fossil fuel, armaments, gambling, property, pharmaceutical and financial services industries have been documented in a number of countries.[46] In each case, the tactics of state capture are drawn from financial donations, lobbying, revolving door jobs, repurposing key institutions, biased research and policymaking, and campaigns in the mainstream and social media. These tactics are used in campaigns by individual corporations, industry peak organisations and self-styled 'think tanks'[47] such as the Heartland Institute and the Breakthrough Institute in the USA, the Global Warming Policy Foundation and the Institute of Economic Affairs in the United Kingdom, and the Institute of Public Affairs and the Australian Strategic Policy Institute in Australia. Therefore, an effective strategy for transitioning to the Sustainable Civilisation would unite community groups on the following strategy to weaken and, if possible, eliminate most of these tactics.

Reduce Influence of Financial Donations and Expenditure

Individual donations to registered political parties should be capped at a very low level, such as $1000 or 1000 euros per person per year, and must be declared immediately on a publicly accessible website. The donor must be clearly identified. Cash donations must be illegal, as must all direct donations to politicians except Independents, who are individual Parliamentarians who do not belong to a political party. Strict limits must be placed on amounts that anyone can spend on influencing an election by advertising. Even better, would be to ban election advertising entirely, with information about candidates provided by an independent organisation. Ideally, elections should be publicly funded, but this could exclude new parties and independents unless special provisions could be made for them. It requires very strong public pressure to induce politicians to implement rules that restrain themselves.

Create Institutions to Monitor Integrity

- Anti-corruption or integrity commissions should be set up for each national and state/provincial sphere of government with guaranteed annual funding that cannot be removed or reduced by the reigning government.
- A national audit office to review government spending is required with a similar funding arrangement.
- Politicians who clearly and deliberately mislead the public, as judged by the integrity commission, should be penalised.
- A public register of lobbyists should be created and work diaries of all elected politicians—especially presidents, prime ministers, ministers and parliamentary or congressional committee chairs—should be published.
- A Commission for the Future, which reports to Parliament, or Congress in the case of the USA, should be created to (i) provide a wider and more accurate basis for political decision-making and (ii) promote public awareness and stimulate the process of increasing public participation in the debate on possible futures. Previous models were Sweden's Secretariat for Futures Studies (1973–1987)[48] and its successor, the Institute for Futures Studies (1987 onwards); New Zealand's Commission for the Future (1977–1982) and its successors[49]; and Australia's Commission for the Future (1986–1998).[50]

Limit Revolving Door Jobs

Ministers, committee chairs, other powerful politicians and their advisors should have a mandatory cooling-off period of at least three years after leaving politics before they may accept an appointment in an industry for which they were responsible while in politics. Furthermore, the process of appointing advisors to politicians in government should be merit-based. Another option is to limit politicians to one or two terms in office, reducing the opportunities for them to be captured.

Require Decision-Makers to Declare Assets

Public officials and politicians should be required to declare annually their assets, liabilities, income from all sources, gifts, advantages and other benefits, as well as unpaid contracts and employment, unpaid board memberships and directorships, participation in organisations, trade unions and NGOs.

Restrict Media Ownership and Control Social Media

Large fractions of the mass media and social media have been captured by vested interests who deny climate science and promote junk science, campaign for delays to climate action and attack high-profile individuals working to publicise the climate crisis.[51] More generally, vested interests use the media and social media to promote neoliberal economics, excessive consumption, social divisiveness and undemocratic governance. The situation is exacerbated by extensive media ownership by a single corporation in the nation, a state/province or a major city. Media ownership should be limited by law. The even more difficult task, to place controls on social media to reduce lies, misrepresentations, defamations and biased searches, must also be tackled. Policy recommendations have been made by the Institute for Strategic Dialogue.[52]

Reduce Legal Power of Corporations

In many jurisdictions, a company has the legal capacity and powers of an individual, as well as all the powers of a body corporate.[53] This situation needs review, with the goal of reducing powers that impinge on the rights of citizens. In particular, corporate rights to acquire or damage land against the wishes of owners and traditional Indigenous custodians should be removed.

Benefits for Democracy

The above recommendations would create systems of government that are fairer and have greater public participation in decision-making, thus giving governments a wider, less biased basis for policies than they receive from vested interests. In other words, they would enable nation-states to operate more democratically.

The recommendations would complement the ongoing growth of grassroots community initiatives—such as local renewable energy projects, community gardens, cooperative organic food stores, cooperative housing, and volunteer environmental protection and remediation groups—in 'mending democracy'.[54] While it may be more difficult to weaken state capture in countries with less established institutions of liberal democracy, especially in autocracies and oligarchies, such governments often take even less interest in the wellbeing of the majority of people. Hence, the role of grassroots community initiatives could be even more important.

6.5   International Agreements

The recommendations to weaken state capture in Sect. 6.4 are relevant to national and sub-national levels and, in a few cases such as the European Union, to a regional level. However, they cannot be applied directly to sustainable development on a global scale. To supplement national strategies and policies, we must urgently improve the Sustainable Development Goals (SDGs), create several new international agreements and strengthen several existing agreements. International agreements can assist in eroding the corporate capture of nation-states of both the Global South and Global North while simultaneously addressing specific environmental and resource conservation issues.

The SDGs are a good set of goals, apart from the absence of limits to population and limits to growth, as discussed in Sect. 3.2. Limits to population are resisted by some politically powerful religious organisations, notably the Roman Catholic Church, which appears to have captured

United Nations reports on population. As a result of an intense campaign by big business,[55] limits to economic growth are regarded as economic heresy by the governments and public services of almost all nation-states. The almost-universal dominance of this neoclassical economics dogma goes beyond state capture to global capture. Correcting the shortcomings of the SDGs must be just the first step in addressing two key drivers of environmental impact, namely growth in consumption and growth in population.

Existing multilateral international agreements comprise treaties, framework conventions and protocols. In the language of the United Nations, a framework convention or agreement describes *a type of legally binding treaty which establishes broad commitments for its parties and leaves the setting of specific targets either to subsequent more detailed agreements (usually called protocols) or to national legislation.* A selection of existing multilateral agreements relevant to sustainable development is listed in Table 6.1.

Although many of the treaties listed in Table 6.1 have been ratified by most or all UN members, progress has been modest in almost all of them. Despite continuing efforts by some governments, by some businesses and by many NGOs, climate change, biodiversity loss, desertification, and proliferation of nuclear weapons (or the capacity to produce them rapidly) have increased over the past two decades. On the positive side, ozone-depleting chemicals are gradually being phased out and the 'hole' in the ozone layer is no longer increasing in size. The general problem is that treaties and conventions are broad, non-enforceable commitments; protocols, although more detailed, are non-enforceable. Despite their limitations, existing global agreements are of some value because they offer precedents and international ethical standards.

International treaties can be used in legal actions by community-based NGOs within some countries. For example, in the Netherlands, a community group, Urgenda, alleged in a class action that the government's plans to cut GHG emissions for all agencies under its control were inadequate. The claim was upheld in a District Court on the basis of the government's general duty of care to its citizens. The government appealed and the Hoge Raad (Supreme Court) upheld the original decision, but on different grounds, namely Articles 2 and 8 of the European Convention

Table 6.1 Selected United Nations treaties, conventions, protocols and other agreements

Name	Parties; comment
Universal declaration on human rights	Adopted in 1948 by the general assembly with 8 nations abstaining but none dissenting
Framework convention on climate change	197 parties
Kyoto protocol	192 parties, not ratified by USA
Paris agreement on climate change	193 parties (USA joined 2016, withdrew 2019 & re-joined 2021)
Convention on biological diversity	All UN members except USA
Convention to combat desertification	197 parties
Vienna convention for the protection of the ozone layer	198 parties
Convention on long-range transboundary air pollution	51 parties
Montreal protocol on substances that deplete the ozone layer	198 parties
Treaty on the non-proliferation of nuclear weapons (aka non-proliferation treaty)	59 parties; four nations have not signed: India, Israel, Pakistan and South Sudan; North Korea withdrew
Comprehensive nuclear test ban treaty	170 Parties; not yet in force as of February 2022
Convention on the prohibition of the use, stockpiling, production and transfer of anti-personnel mines and on their destruction	164 parties
Treaty on the prohibition of nuclear weapons	Adopted July 2017; entered into force in January 2021; as of 26 September 2022, 68 states have ratified or acceded

Sources: Multilateral Treaties Deposited with the Secretary-General (United Nations. https://treaties.un.org/Pages/ParticipationStatus.aspx?clang=_en and https://www.un.org/en/about-us/udhr/history-of-the-declaration)

Notes: 1. A Party is a State with treaty-making capacity that has expressed its consent to be bound by that treaty by an act of ratification, acceptance, approval or accession. Signature without ratification does not make a Party

2. In addition to the treaties listed above, the Antarctic Treaty System, not a United Nations treaty, has 12 Parties, whose scientists had been active in and around Antarctica, and 42 acceding countries. Its provisions include freedom of scientific research, putting aside nations' territorial claims, and non-militarisation of Antarctica. See https://www.ats.aq

of Human Rights. The Court found that the government was required *"to take suitable measures if a real and immediate risk to people's lives or welfare existed and the state was aware of that risk"* and that those preconditions were fulfilled. It also stated that, under the UNFCCC and the Paris Agreement, a Party could not excuse its small target on the grounds that other countries were emitting more emissions.[56]

The need for international agreements on global limits on energy and resource use, pollution and population follows from the fact that Earth's physical resources and its capacity to absorb pollution are finite and, as discussed in Chap. 2, human activities are already exceeding several limits. The argument that science cannot yet quantify all the limits can be countered by reference to the Precautionary Principle: *"Where there are threats of serious or irreversible environmental damage, lack of full scientific certainty should not be used as a reason for postponing measures to prevent environmental degradation"*. In addition to limits, the agreements must also specify a fair distribution of the agreed remaining budgets for GHGs, energy, resources and emissions.

Next, we consider several specific issues where international agreements are needed urgently in addition to national.

Multinational Corporations

Many large multinational corporations are wealthier and have more political power than some countries. International agreements are needed to control the operations of multinational corporations that evade national environmental, social and economic (e.g. taxes) responsibilities by registering in countries with essentially no controls.

Climate Mitigation

A new international agreement is needed to cap global carbon emissions and establish a fair allocation of the remaining emissions between countries.

To keep global heating below a desired temperature increase, such as 1.5 °C or 2 °C, only a limited amount of global fossil fuel resources can be burned, producing a limited amount of emissions. The latter is called the carbon budget for that temperature increase. Once the budget has been emitted, any additional emissions must be net zero, that is, they must be cancelled by removal of GHGs from the atmosphere. At present, no technology for removal is commercially available.[57] Estimates of the remaining carbon budget vary, according to method, the desired probability of meeting the target, the GHGs included in the calculation, and accounting for feedbacks such as the melting of permafrost. Climate scientist Will Steffen estimates that the remaining carbon budget from the beginning of 2020 for global heating of 1.5 °C with 66% probability to be 37 gigatonnes (Gt) of carbon or 136 Gt of CO_2-equivalent (CO_2-e) emissions.[58] This corresponds to less than four years of emissions at the current rate.[59] If you are reading this book in 2023, three of those years have gone. Steffen concludes, and we sadly agree, that *the lower Paris target of 1.5 °C target is now out of reach. Past inaction … [has] cost us dearly*.[60]

Changing the target to 2 °C with 66% probability gives humanity a little more time—the 2 °C budget is about three times[61] the 1.5 °C budget, let's say about 408 Gt CO_2-e from the beginning of 2023. If the pre-COVID rate of emissions is maintained, that carbon budget would be exhausted before 2033.[62] To say that we are facing a climate emergency is not an exaggeration!

What is a fair way to divide the remaining carbon budget between countries? One way, that would be consistent with the Universal Declaration on Human Rights, would be to allocate each person the same amount: 408 Gt of emissions divided by eight billion people gives 51 tonnes per person. This is *not* 51 tonnes per year but 51 tonnes forever! Table 6.2 shows the recent pre-COVID emissions per person per year in selected countries and the number of years they could continue to emit at that rate, counting from the beginning of 2023. The table only shows the territorial emissions for each country. For rich countries, these are generally exceeded by the emissions embodied in net imports. Rapidly growing economies, such as China and India, produce most of the manufactured goods for the rich countries.

Clearly, to keep global heating below 2 °C, while allowing low-income countries to develop, the rich countries must greatly reduce their emissions. China too must stabilise and reduce its emissions to the global

Table 6.2 Average GHG emissions per person per year and remaining years at that rate

Country	Average emissions per person in 2019 (tonnes of CO2-e)	Remaining years of budget at this emissions rate, assuming equal amounts per person in the world
Australia	19	2.8
USA	15	3.5
European union	8.25	6.3
China	7.4	7.0
India	2.2	23

Source: Our calculations are based on 2019 GHG emissions by country (Hannah Ritchie & Max Roser. Greenhouse gas emissions. https://ourworldindata.org/greenhouse-gas-emissions) and population by country (United Nations. https://population.un.org/dataportal/home)
Notes: 1. Approximate values, due to uncertainties in non-CO_2 emissions
2. Carbon budget for 2°C commencing 1/1/2023

average. It could do this if the rich countries changed their imported goods from China to those with less embodied carbon emissions. One policy option would an international agreement to cap the remaining global carbon budget at 408 Gt total and divide it according to population. But, by the time such an agreement could be reached, many of the rich countries would have already exceeded their fair share of the rapidly shrinking budget. An alternative policy approach, which could possibly be negotiated more rapidly, would be an international emissions trading scheme whose cap is the remaining carbon budget for the agreed temperature increase. It would be difficult to reach international agreement on this while the governments of two of the biggest emitters per capita, the USA and Australia, reject any kind of carbon price.

A new international agreement that is needed urgently is a Fossil Fuel Non-Proliferation Treaty.[63] A step in this direction was taken at the 26th Conference of the Parties (COP26) to the UN Framework Convention on Climate Change with the formation of the Powering Past Coal Alliance.

International Caps on Use of Renewable Resources

Responsibility for our planet's soils, forests, and indeed all biodiversity cannot be limited to national and sub-national governments. The continuing destruction of the Amazon rainforest contributes to global climate change, reduces oxygen levels in the air we breathe and diminishes the available ingredients for medicines. The continuing loss of soils from destructive agricultural practices, erosion and climate change in the USA[64] and Australia reduces global food production. Therefore, international agreements are needed to supplement national policies for resource conservation and, in some countries, to exert pressure on national governments to implement stronger national policies. These agreements must be designed to constrain the power of multinational agriculture, fisheries and forestry industries. The rates at which we humans extract renewable resources from nature must be reduced to the rates at which they are replenished.

However, renewable energy (RE) has to be limited more severely, because a global RE system that replaced all current fossil fuel use would need only a tiny fraction of the solar energy arriving on Earth's surface. If the rate of energy consumption were permitted to grow to the rate at which RE is replenished by the Sun, it could reach thousands of times its current level. Then, so would the size of the global economy with all its environmental impacts. Thus, while the direct environmental impacts of most RE sources are very small, the indirect environmental impacts of very large amounts of RE production would be devastating. Global energy consumption, and hence energy consumption by the rich countries, must be reduced, as discussed in Sect. 4.5.

International Agreement on Use of Non-renewable Resources

While national policies could limit the use of scarce raw materials through design of products for disassembly, reuse, remanufacturing, recycling, substitution, and improved efficiency of manufacturing (see Chap. 5), international agreements are needed to place global caps on extraction and trade of these resources and to constrain the multinational mining industry. The priority must be for resources that are rare and needed for the transition to the Sustainable Civilisation.

Reducing the Incidence of War and the Likelihood of Nuclear War

Nine countries—China, France, India, Israel, North Korea, Pakistan, Russia, the United Kingdom, and the United States—currently have nuclear weapons. The Nuclear Non-Proliferation Treaty (NPT)[65] entered into force in 1970. It aims to stop the spread of nuclear weapons to non-nuclear weapons states (Articles IIII) and undertake the nuclear disarmament of nuclear weapons states (Article VI). However, little if any progress has been made on the latter. South Africa is the only country to develop nuclear weapons and then relinquish them. Ukraine had nuclear weapons left over from the collapse of the former Soviet Union in 1991, but returned them to Russia in 1996.

The NPT may have slowed nuclear weapons' proliferation, but has not stopped it. Three countries—Israel, India, and Pakistan—never joined the NPT, and North Korea left the treaty in 2003. South Africa acceded to the Treaty in 1991, two years after it had ended its nuclear weapons program. India, Pakistan, North Korea and South Africa developed their nuclear weapons under the cloak of 'peaceful' nuclear energy. The UK expanded its nuclear weapons stock with plutonium reprocessed from the waste of its first generation of nuclear power stations. In France, the civil and military nuclear industries overlap.[66]

The NPT must be amended to place under complete international control the two stages of the nuclear fuel cycle that can lead to the production of nuclear explosives, namely uranium enrichment and reprocessing of spent fuel. These explosives are highly enriched uranium, Uranium-233 produced from thorium, and Plutonium-239 produced from the fission of Uranium-235. At present, under the NPT's Article III, each Non-Nuclear Weapon State is required to conclude a safeguards agreement with the International Atomic Energy Agency (IAEA). This involves inspection of facilities, but not control.

Additional Parties to the new Treaty on the Prohibition of Nuclear Weapons[67] are needed to expand pressure on nuclear weapons powers to denuclearise. At present, governments of countries with their own nuclear weapons, or have nuclear weapons of other countries stationed on their land, are not Parties. So far, only one European country, Austria, is a Party. Campaigns are needed to pressure non-nuclear countries that are not Parties to join or accede to the treaty. As of October 2022, non-nuclear weapons countries that are not Parties include Argentina, Australia, Canada, Egypt, Morocco, Singapore, Papua New Guinea, Tunisia and Turkey. Meanwhile, communities in all nuclear weapons states should pressure their governments to declare and legislate that they will not be the first to use nuclear weapons.

Decisions to take a country to war are sometimes made by governments for unjustifiable reasons, such as the desire to capture another country's natural resources or to increase the power of the ruling elite of the aggressor country or even to build voter support for a forthcoming election. In several countries that are nominally democratic, such as Australia Russia, and the United Kingdom, the power to make war is held by the Prime Minister, President or Cabinet alone, without any requirement for Parliamentary debate and approval. This power must be curbed. Most people do not want wars, so campaigns to strengthen democracy, by community participation in national and sub-national decision-making, can reduce the probability of governments making war.[68]

Social defence—otherwise known as nonviolent defence, civilian-based defence and defence by civil resistance—is an alternative to military defence that can sometimes be effective with fewer casualties. "*Social defence is nonviolent community resistance to repression and aggression, as an alternative to military forces. 'Nonviolent' means using rallies, strikes, boycotts and other such methods that do not involve physical violence against others.*"[69] Nonviolent community resistance has been used successfully, or partially successfully, in a number of cases[70]:

- It ended the Kapp putsch, which was an attempted coup by the military in Germany in 1920.
- When the Soviet Union invaded of Czechoslovakia in 1968, spontaneous, mostly nonviolent, citizen resistance delayed the establishment of a puppet government for eight months and undermined the credibility of the Soviet Communist Party.

- In 1998 there was a political coup, supported by the military and the KGB, against the Russian leader, Mikhail Gorbachev. His policies were regarded as too liberal by the old guard. The citizens refused to cooperate through protests, strikes, rallies and circulating illegal newsletters. They chatted and joked with the bewildered troops. Within a few days, the coup collapsed.

These cases of nonviolent resistance were conducted spontaneously, without the training and organisation of social defence. Another precondition for successful social defence is strong local communities. Therefore, grassroots projects such as community renewable energy, organic food cooperatives, community childcare and community radio can all contribute indirectly to social defence. It could play a stronger role in resistance to potential external aggression if governments included it in their defence strategy. But governments may be reluctant to foster it because the people can also use it against repression by their own governments.

Sovereign Debt Relief for the Global South

Out of 148 countries in the Global South, 135 are 'critically indebted'. In many of these countries, servicing the debt can only be maintained at the expense of public services such health, education and social security. Climate mitigation and adaptation, and implementing the Sustainable Development Goals, are also given lower priority than debt servicing. The situation has deteriorated substantially since before the COVID-19 pandemic, for example, as the result of loss of tourism revenue in Small Island Developing States.[71]

The increased prices of fossil fuels and grains resulting from the war in Ukraine are further exacerbating the debt situation for many countries.[72] Although the G20[73] launched a debt moratorium during the pandemic, it only gave a small amount of relief, that ended in December 2021, to 48 low-income countries. Private creditors, who hold the majority of the debt, did not participate.[74] In summary, *"the current model of development finance built on market-based mechanisms is fundamentally broken"*.[75]

Solutions proposed by NGOs include[76]:

- for the poorest countries that have no capacity to repay, the cancellation of debt or its conversion into development finance
- for other countries, the renewal and extension of the G20 debt moratorium for a longer period and to more countries
- the inclusion of private sector creditors in debt restructuring negotiations within the G7, G20 and International Monetary Fund
- an international agreement that debtor countries may default on payments in their dealings with private creditors that refuse to cooperate in debt restructuring
- an International Economic Reconstruction and Systemic Reform Summit under the auspices of the UN.

These solutions can be justified ethically by recognising that the economic growth of the Global North has relied, and still relies to some extent, on the extraction of resources from the Global South.

International People-to-People Institutions

Agreements under the auspices of the UN are made by governments, which do not always act in the interests of their peoples or the planet. Even when they do try to act in these interests, governments often have great difficulties in reaching effective intergovernmental agreements of substance. Therefore, there could be an important potential role for citizen-based global institutions to sit in parallel with those of the UN. Such institutions can be built up by strengthening international links between national community organisations dedicated to environmental protection, social justice, peace and public health. The fortified organisations can exert pressure for change in both international arenas and on individual recalcitrant governments. Inspiring models include the International Campaign to Abolish Nuclear Weapons (ICAN), upon whose initiative and efforts the United Nations Treaty on the Prohibition of Nuclear Weapons was set up,[77] the international School Strike for Climate movement inspired and empowered by Greta Thunberg,[78] and the International Baby Food Action Network (IBFAN).[79]

6.6   Conclusion

Many nation-states have been captured by corporate and other vested interests using tactics such as political donations, lobbying, 'revolving door' jobs, repurposing institutions, astroturfing, biased research, policy-making, bribery and corruption, and public influence campaigns in the media and social media. These tactics have enabled vested interests to determine the policies of governments, opposition parties and public officials on a wide range of environmental, social and economic issues impacting ecologically sustainable and socially just development. Community pressure to weaken these tactics can have multiple benefits, supplementing campaigns on single issues. Recommended campaign goals are summarised in Box 6.3. The over-arching campaigns can bring together many community organisations to exert much greater social and political pressure for change than many small groups acting independently on single issues.

Box 6.3 Goals for Over-arching National and Sub-national Campaigns

- Expose and reduce the influence of political donations and political expenditure.
- Create institutions to monitor integrity of politicians, political advisors and public officials.
- Limit revolving door jobs.
- Limit media ownership.
- Reduce the legal power of corporations.
- Ensure greater transparency and community participation in government decision-making.

We also need international agreements to complement national and sub-national policies. These agreements would cap global emissions, implement a fair distribution of the remaining global carbon budget and other scarce resources to all people, reduce the rate of use of renewable resources to the replenishment rate (and well below that for energy), conserve non-renewable resources and, where possible, replace them with renewable resources, improve social justice, reduce the frequency of wars, alleviate sovereign debt of the Global South and, last but not least, limit the power and influence of multinational corporations.

Notes

1. Suzy Kassem (2011). *Rise Up and Salute the Sun: The writings of Suzy Kassem*. Awakened Press.
2. Nick Ritchie (2013). Trident in UK politics and public opinion. https://basicint.org/publications/dr-nick-ritchie/2013/trident-uk-politics-and-public-opinion; Harry Hayball (2016). Does the British public support Trident? Opinion polls say no. https://www.stopwar.org.uk/article/does-the-british-public-support-trident.
3. British Pugwash (2021). https://britishpugwash.org/uk-public-opinion-on-first-use-of-nuclear-weapons-oct-2021.
4. Murray Goot (2003). Public opinion and the democratic deficit: Australia and the war against Iraq. *Australian Humanities Review* Issue 29, May. http://australianhumanitiesreview.org/2003/05/01/public-opinion-and-the-democratic-deficit-australia-and-the-war-against-iraq.
5. Nick O'Malley & Miki Perkins (2021). *Sydney Morning Herald,* 30 August. https://www.smh.com.au/environment/climate-change/australia-s-biggest-climate-poll-shows-support-for-action-in-every-seat-20210829-p58mwb.html.
6. Greg Jericho (2021). *Guardian Australia.* https://www.theguardian.com/business/grogonomics/2021/may/25/australia-has-not-had-a-gas-led-recovery-not-in-jobs-not-in-tax-receipts.
7. Jake Evans (2022). *ABC News.* https://www.abc.net.au/news/2022-08-03/43-pc-emissions-reduction-target-to-become-law-greens-support/101295224.
8. Right Livelihood. https://rightlivelihood.org/what-we-do/the-right-livelihood-award/about-the-award.
9. These drives are not entirely autonomous psychological processes. They can be controlled to some degree by social institutions. The legal system and other institutions are intended to limit such power, but in practice are often captured by the powerful.
10. Michael Webber (2022). China: capitalism and change?. In: Samuel Alexander et al. (eds). *Post-Capitalist Futures: Paradigms, politics, and prospects*. Palgrave Macmillan, Chapter 6.

11. Astroturfing is the attempt to create the false impression of widespread grassroots support for a policy, individual, or product.

12. Ivor Chipkin et al. (2018). *Shadow State: The politics of state capture*. Wits University Press, https://doi.org/10.18772/22018062125.

13. Australian Democracy Network (2022). *Confronting State Capture*. https://australiandemocracy.org.au/statecapture.

14. Sharon Beder (n.d.). *Business-Managed Democracy*. https://www.herinst. org/BusinessManagedDemocracy/introduction/index.html.

15. AAP (2003). *The Age*. 17 October, https://www.theage.com.au/national/ bushs-sheriff-comment-causes-a-stir-20031017-gdwk74.html.

16. Scott Bennett & Rob Lundie (2007). Australian Electoral Systems. Research Paper no. 52007–08. https://www.aph.gov.au/About_ Parliament/Parliamentary_Departments/Parliamentary_Library/pubs/ rp/RP0708/08rp05.

17. Centre for Public Integrity (2020). https://publicintegrity.org.au/wp- content/uploads/2021/02/Briefing-paper-Hidden-money-2020.pdf.

18. ABC 4 Corners program. https://www.abc.net.au/news/2019-11-01/ scott-morrison-environmental-groups-targeting-businesses- boycott/11660698.

19. Guy Pearse (2007). *High and Dry: John Howard, climate change and the selling of Australia's future*. Penguin/Viking.

20. Robert Manne (2011). Rudd's downfall: written in The Australian. https://www.abc.net.au/news/2011-09-05/manne-rudds-downfall-written- in-australian/2869942.

21. Australian Democracy Network, *op. cit.*

22. Greg Bailey (2016). The Liberal Party and the Institute of Public Affairs: Who is Whose? *Pearls and Irritations*, 1 April. https://johnmenadue. com/greg-bailey-the-liberal-party-and-the-institute-of-public- affairs-who-is-whose.

23. The IPA's original 75 policy demands, published in 2012, on a future Coalition government: https://ipa.org.au/ipa-review-articles/be-like- gough-75-radical-ideas-to-transform-australia; critical comment in 2013 following election of the Coalition: https://www.crikey.com. au/2013/09/06/institute-of-liberal-party-policy-what-the-ipa-will-get- from-abbott/. After the government had implemented many of the policy demands, the IPA issued supplementary demands in 2019: https:// ipa.org.au/wp-content/uploads/2019/04/IPA-Research-20-Policies-to- Fix-Australia.pdf.

24. Michael Mazengarb (2021). *Renew Economy*, 18 February. https://reneweconomy.com.au/nationals-push-nuclear-in-new-attempt-to-highjack-cefc-changes/; Michael Mazengarb (2022). *Renew Economy*, 28 March, https://reneweconomy.com.au/senate-again-blocks-angus-taylors-bid-to-redirect-arena-funds-to-ccs-projects/.

25. Michael Mazengarb (2020). *Renew Economy*, 11 August. https://reneweconomy.com.au/covid-commission-advised-morrison-to-underwrite-gas-pipelines-but-ignored-green-jobs-48758.

26. Nour Haydar (2019). *ABC News*, 1 November. https://www.abc.net.au/news/2019-11-01/scott-morrison-environmental-groups-targeting-businesses-boycott/11660698.

27. Oliver Milman (2013). *The Guardian Australia.* https://www.theguardian.com/environment/2013/oct/30/one-third-of-australias-media-coverage-rejects-climate-science-study-finds.

28. Ketan Joshi (2020). *RenewEconomy*, 21 December. https://reneweconomy.com.au/news-corp-has-caused-massive-climate-delay-but-its-grip-on-power-is-slipping-97148.

29. Chipkin et al. (2018), *op. cit.*; Centre for Public Integrity (2020), *op. cit.*; Lucas (2021), *op. cit.*; Australian Democracy Network (2022), *op.cit.*

30. Adam Lucas (2021). Investigating networks of corporate influence on government decision-making: the case of Australia's climate change and energy policies. *Energy Research & Social Science* 81:102271. https://doi.org/10.1016/j.erss.2021.102271.

31. Pieter Wezeman et al. (2021) Trends in international arms transfers. SIPRI Fact Sheet. https://sipri.org/sites/default/files/2021-03/fs_2103_at_2020.pdf.

32. Australian Strategic Policy Institute. https://www.aspi.org.au/about-aspi/funding.

33. The White House (2021). https://www.whitehouse.gov/briefing-room/statements-releases/2021/09/15/joint-leaders-statement-on-aukus/.

34. Richard Broinowski (2022). *Fact or Fission: The truth about Australia's nuclear ambitions.* 2nd edition. Scribe.

35. Craig Whitlock and Nate Jones (2022). Former U.S. navy leaders profited from overlapping interests on sub deal. *Washington Post* 18 October. https://www.washingtonpost.com/investigations/interactive/2022/australia-nuclear-submarines-us-admirals.

36. Mike Scrafton (2022). US admirals driving AUKUS has a conflict of interest: Washington Post. *Pearls and Irritations*, 25 October. https://

johnmenadue.com/us-admirals-driving-aukus-had-conflict-of-interestn-washington-post.

37. MichaelWest Media (n.d.). https://www.michaelwest.com.au/air-chief-marshal-mark-binskin-ac-retd.
38. Thales (2020). https://www.thalesgroup.com/en/australia/press-release/duncan-lewis-ao-dsc-csc-join-thales-australia-board.
39. Auditor-General Report No. 6 of 2018–19. https://www.anao.gov.au/work/performance-audit/army-protected-mobility-vehicle-light.
40. Australian Democracy Network, *op. cit.*
41. Christopher Knaus (2019). *The Guardian Australia.* 26 June. https://www.theguardian.com/australia-news/2019/jun/26/christopher-pyne-takes-job-with-consulting-firm-ey-to-help-grow-defence-business.
42. Luciana Dias (2019). Social environmentalism and corporate capture. In: Beate Sjåfjell and Christopher Bruner (eds). *The Cambridge Handbook of Corporate Law, Corporate Governance and Sustainability.* Cambridge University Press, Chapter 25.
43. Greenpeace (2008) *Slaughtering the Amazon.* https://www.greenpeace.org/usa/wp-content/uploads/legacy/Global/usa/planet3/PDFs/slaughtering-the-amazon-part-1.pdf.
44. Lee Fang (2017). Sphere of influence: how American libertarians are remaking Latin American politics. *The Intercept,* 9 August. https://theintercept.com/2017/08/09/atlas-network-alejandro-chafuen-libertarian-think-tank-latin-america-brazil/.
45. Dias, *op. cit.*
46. Transparency International. https://www.transparency.org.
47. Naomi Oreskes & Erik Conway (2010). *Merchants of Doubt: How a handful of scientists obscured the truth on issues from tobacco smoke to global warming.* Bloomsbury, New York.
48. Göran Bäckstrand (1981). Sweden's secretariat for futures studies. *World Futures* 17(3–4):275–297. https://doi.org/10.1080/02604027.1981.9971940.
49. Bob Frame (2018). New Zealand: new futures, new thinking? *Futures* 100:45–55.
50. Richard Slaughter (1999). Lessons from the Australian Commission for the Future: 1986–1998. *Futures* 31:91–98.
51. Jennie King, Lukasz Janulewicz, Francesca Arcostanzo (2022). *Deny, Deceive, Delay: Documenting and responding to climate disinformation at COP26 and beyond: Executive Summary.* Institute for Strategic Dialogue,

https://www.isdglobal.org/wp-content/uploads/2022/06/Executive-Summary-Deny-Deceive-Delay.pdf.

52. Jennie King et al. (2022), *op. cit.*

53. For example, Australia's Corporations Act 2001. http://www5.austlii.edu.au/au/legis/cth/consol_act/ca2001172/s124.html.

54. Carolyn Hendriks, Selen Ercan & John Boswell (2020). *Mending Democracy: Democratic repair in disconnected times.* Oxford University Press.

55. Kerryn Higgs (2014). *Collision Course: Endless growth on a finite planet.* MIT Press.

56. Michael Kirby (2022). Climate change litigation and human rights. In: Stephen Williams & Rod Taylor (eds) *Sustainability and the New Economics: Synthesising Ecological Economics and Modern Monetary Theory.* Springer, Section 13.3.

57. Apart from tree-planting, which would require a huge area of land to absorb a significant amount of global emissions. This measure would be reversed by wildfires.

58. Will Steffen (2022). The Earth system, the Great Acceleration and the Anthropocene. In: Stephen Williams & Rod Taylor (eds) *Sustainability and the New Economics: Synthesising Ecological Economics and Modern Monetary Theory.* Springer, Chapter 2.

59. The recent pre-COVID rate of human CO_2 emissions is about 42 Gt p.a. At this point in the rough calculation we ignore non-CO_2 GHG emissions, because there is a wide range of uncertainty in the size of their contribution.

60. Steffen, *op. cit.*

61. IPCC (2018). *Special Report: Summary for Policymakers. Global Warming of 1.5 °C.* Masson-Delmotte, V. et al. (eds). https://www.ipcc.ch/sr15/chapter/spm/, Table 2.2.

62. Taking account of non-CO_2 GHG emissions would probably bring this to before 2030.

63. Peter Newell & Andrew Simms (2020). Towards a fossil fuel non-proliferation treaty. *Climate Policy* 20:1043–1054. https://doi.org/https://doi.org/10.1080/14693062.2019.1636759; The Fossil Fuel Non-Proliferation Treaty website. https://fossilfueltreaty.org/.

64. Union of Concerned Scientists (2020). https://www.ucsusa.org/about/news/national-soil-erosion-rates-track-repeat-dust-bowl-era-losses-eight-times-over.

65. International Atomic Energy Agency (1970). *Treaty on the non-proliferation of nuclear weapons.* https://www.iaea.org/sites/default/files/publications/documents/infcircs/1970/infcirc140.pdf.

66. See reports by the Nuclear Weapons Archive and the Institute for Science and International Security; and summary in Mark Diesendorf (2014). *Sustainable Energy Solutions for Climate Change.* UNSW Press and Routledge-Earthscan, Chapter 6.

67. United Nations. *Treaty on the prohibition of nuclear weapons.* https://www.un.org/disarmament/wmd/nuclear/tpnw.

68. Brian Martin (1984). *Uprooting War.* Freedom Press. Web edition (1990): https://www.bmartin.cc/pubs/90uw.

69. Jørgen Johansen & Brian Martin (2019). *Social Defence.* Sparsnäs, Sweden: Irene Publishing. Web edition: https://www.bmartin.cc/pubs/19sd/19sd.pdf.

70. Johansen & Martin, *op. cit.*, Chapter 3.

71. Bodo Ellmers et al. (2022). *Global sovereign debt monitor 2022.* Aachen and Düsseldorf: MISEREOR & erlassjahr.de. https://erlassjahr.de/wordpress/wp-content/uploads/2022/04/GSDM22-online.pdf.

72. Ulrich Volz et al. (2022). *Addressing the Debt Crisis in the Global South: Debt relief for sustainable recoveries.* Think7. https://www.think7.org/publication/addressing-the-debt-crisis-in-the-global-south-debt-relief-for-sustainable-recoveries.

73. G20: a group of 20 rich and rapidly growing economies.

74. Volz et al. *op. cit.*

75. Daniel Munevar (2021). *Dynamics and implications of the debt crisis of 2020.* Eurodad. https://assets.nationbuilder.com/eurodad/pages/2112/attachments/original/1622627378/debt-pandemic-FINAL.pdf?1622627378.

76. Ellmers et al. *op. cit.*; Volz et al. *op. cit.*; Debt Justice, https://debtjustice.org.uk; Munevar *op. cit.*

77. ICAN was awarded the 2017 Nobel Peace Prize for this work.

78. Greta Thunberg was awarded the Right Livelihood Award for 2019.

79. IBFAN, https://www.ibfan.org/, was awarded the Right Livelihood Award in 1998.

Reference

Weblinks accessed 27/10/2022.

7

Transforming the Economic System

Virtually every aspect of conventional economic theory is intellectually unsound; virtually every economic policy recommendation is just as likely to do general harm as it is to lead to the general good. Far from holding the intellectual high ground, economics rests on foundations of quicksand.
(Steve Keen[1])

Much popular understanding and political rhetoric concerning national economic policies are based on myths, which some critics call 'lies'. It's a myth that the government of a rich country must cut the wages of workers and taxes to remain 'competitive'; that citizens must accept increasing inequality so that big business can create jobs for the masses; that the government must privatise government services to be 'efficient'; and that the government of a rich country cannot afford to ensure that teachers, child-care and aged-care workers, and pensioners have adequate income. Governments have choices about allocating money. The myths serve the vested interests that dominate the choices of socioeconomic policies by many governments.[2]

The above myths and others have been exposed and refuted by economist Richard Denniss within the framework of conventional economics.[3] But that framework is itself inadequate for creating ecologically

© The Author(s), under exclusive license to Springer Nature Singapore Pte Ltd. 2023
M. Diesendorf, R. Taylor, *The Path to a Sustainable Civilisation*,
https://doi.org/10.1007/978-981-99-0663-5_7

sustainable and socially just societies. Conventional economics is built upon flawed foundations, on assumptions at odds with the real world. Of particular concern in this chapter, it is based on the incorrect notion that endless growth on a finite planet is possible. Therefore, we must replace conventional economics with a new economics that facilitates a transition to a civilisation consistent with human and planetary wellbeing.

This chapter summarises the reasons why conventional economics has failed to protect the environment and social justice (Sect. 7.1) and then, in Sect. 7.2, introduces the interdisciplinary field of ecological economics, that offers a broad framework for a sustainable society. Section 7.3 discusses the need for planned degrowth by the rich countries and Sect. 7.4 explains how rich countries that have monetary sovereignty can finance a global transition to the Sustainable Civilisation. Section 7.5 proposes a strategy for transitioning the socioeconomic system.

7.1   The Failure of Conventional Economics

When people refer to 'conventional economics', they mean neoclassical economics, which is the dominant form of economics today. Calling it 'conventional' reflects its take-over of economic thinking, despite its inadequacies outlined in this section. Its practitioners have given it status as a science by dressing it up to imitate physics[4] and giving it a mathematical structure. They even created a 'Nobel' prize for it, although there is no Nobel Prize in economics; it's actually the Sveriges Riksbank (Bank of Sweden) Prize in Economic Sciences in Memory of Alfred Nobel.[5]

Neoclassical economics is a theoretical framework that focuses on supply and demand as the driving forces behind the production, pricing and consumption of goods and services. It assumes that people are simply self-interested consumers who compete for 'goods' and services. But, in the real world, people cooperate as well as compete, as witness the existence of many business cooperatives, community organisations and institutions of government, yet such organisations are excluded from neoclassical economics theory.

Neoclassical economics further assumes that individual economic actors have rational preferences and perfect knowledge of the market. It assumes that they seek to maximise a poorly defined concept called 'utility', the total satisfaction or welfare derived from consuming a good or service. Unlike a property of a physical system, utility cannot be measured directly, but only by surveys of consumers. The utility of an item is subjective, varying with the group surveyed, income, wealth, social class, cost of item, advertising, geographic location, time, and so on. It is measured in 'psychological' units. It is based on preferences for market goods and services only, with the implicit assumption that non-market goods and services contribute little to satisfaction or welfare. Humans are assumed to be insatiable, meaning that they can never have enough of all goods and services. This leads to the dangerous belief that endless economic growth is natural and desirable.

As an imitation of physics, neoclassical economics has invented its own 'forces', so-called market forces, and systems that always trend towards equilibrium under the action of these 'forces'. Neoclassical economic systems are assumed to be either in equilibrium, where supply balances demand, or near equilibrium. For mathematical convenience, the economic system is generally assumed to be a 'linear system', meaning that its variables are proportional to one another—a small change in one variable produces small changes in other variables.

In reality, the 'forces' in neoclassical economics are metaphorical ones, having nothing in common with physical forces. Moreover, a real economic system is generally a dynamic system, not static, and, over time, it does not necessarily approach equilibrium at all. It experiences trade or business cycles, including boom and bust events, that are far from equilibrium—the Global Financial Crisis of 2007–2008 is just one of many examples. Therefore, many real economic systems cannot be validly modelled by equilibrium or near-equilibrium systems with linear relationships between their variables.[6] *It follows that much of the impressive-looking mathematical modelling in macroeconomics is meaningless.* Yet the mathematical structure of neoclassical economics helps to give it undeserved scientific status, making it difficult to understand by intelligent, well-educated people lacking a mathematical background. This source of status is undeserved, because its mathematical models have little relationship to actual economic systems comprising real people and real institutions.

Some of the most unrealistic macroeconomic models are on climate mitigation in the energy sector. Many assume, contrary to much empirical evidence, that markets for all energy services are competitive and therefore there are no energy efficiency options still to be implemented that have zero net costs; they estimate the costs but rarely the benefits of mitigation; they substitute dubiously derived parameters for specifications of technologies; they assume linear relationships; and they often fail to perform sensitivity analyses, that is, to explore the extent to which the results depend on particular assumptions.[7] By excluding the risk of climate tipping points, which are non-linear events, from occurring in the twenty-first century, neoclassical economist William Nordhaus has published studies[8] that minimise and trivialise the cost of climate impacts. As these studies are inconsistent with climate science as well as real economies, they are doubly invalid.[9]

The inadequate treatment of climate change, many other environmental issues and social justice by neoclassical economics suggests that Steve Keen (quoted at the beginning of the chapter) may have been too kind, and that neoclassical economics could be doing *more* harm than good.

This is not surprising, because the neoclassical economic system is separate from the natural environment, although it recognises limited links. Environmental and resource economics, branches of neoclassical economics, treat the environment as external to the economy while recognising that the environment fulfils three limited roles: resources, a waste dump and amenities. This conception lacks understanding that the economy is completely dependent on the existence and good health of the environment, that the economy is a subset of society, and that society is in turn a subset of the biosphere (Fig. 3.1). It rejects the reality that a finite planet has limited resources and limited capacity to absorb wastes.

Although environmental economics makes some economists more aware of the importance of the environment, which is a good thing, it retains the limitations inherent in neoclassical economics: in particular, it's assumed that markets handle 'externalities' such as pollution.[10] To mitigate human-induced climate change, it relies on a carbon price and prefers to apply this via emissions trading schemes (ETS). However, in the real world where markets are generally imperfect, pricing is insufficient as a mitigation policy. For example, people who live in rental accommodation are limited in the extent they can make their homes more energy efficient, whatever the carbon price. People who live on an urban fringe with inadequate public transport are forced to drive their cars, whatever the carbon price. Strategic long-term planning by government is needed to build the infrastructure (e.g. new long-distance transmission lines; new railway lines) for a sustainable energy future—the market does not build it automatically.[11] Nevertheless, a carbon price can contribute to climate mitigation as part of a portfolio of climate response policies. A carbon tax reduced greenhouse gas (GHG) emissions from the Australian National Electricity Market during the short period (mid-2012 to mid-2014) of its existence[12] and the European Union's ETS has contributed in recent years to the reduction in its emissions from stationary energy,[13] despite failure in its early phases.[14]

In neoclassical economics, the concept of economic sustainability involves maintaining consumption at an 'adequate' level into the future, by substituting human capital for natural resources. Thus, according to this flawed logic, a new suburb can substitute for a forest; an increase in agriculture in Siberia as the permafrost melts can substitute for a stable climate; and intensively farmed fish laden with antibiotics can substitute for wild fish. However, there is no substitute for phosphorus, an essential nutrient for plants and humans (see Chap. 5).

The inevitable outcome of those substitutions that are technically possible is the destruction of our life support system, the biosphere, and, as a consequence, the economic system as well. In the ecological economics literature, this concept of economic sustainability is regarded as 'weak', that is, not 'strong' in the sense discussed in Sect. 3.1, because it assumes trade-offs between the environment and the economy. However, a strong definition of sustainable development requires types of economic and social development that protect and restore the natural environmental and social justice.

Neoclassical economics assumes that poverty will be overcome by economic growth. The usual slogans are "a rising tide lifts all ships" and "a bigger cake gives more for all". These slogans assume that wealth trickles down from the rich to the poor. In a few cases this happens, but in most cases, it does not. Huge inequalities in wealth and income remain (see Sect. 2.2).

A key link between ecological sustainability and social justice is that consumption by the world's richest citizens is responsible for the vast majority of environmental impacts.[15] On top of direct consumption, their savings and investments cause additional environmental impacts.

A logical response to this environmentally destructive and socially unjust situation would be to limit the wealth and income of the rich and redistribute some of the excess to the poor. However, neoclassical economists and their disciples seem to believe that this is politically impossible and therefore that continuing economic growth, even in the rich countries, is the only solution.[16]

A further blow to social justice has been struck by the rise of *neoliberal economics*, also known as *neoliberalism*. This is an economic ideology that borrows from the theoretical assumptions of neoclassical economics and, more generally, from free-market capitalism. Its central theme is the notion that markets, especially financial markets, generally outperform governments in the allocation of resources and investments.[17] Hence its proponents campaign for privatisation, free trade, low taxes, low regulation and low government spending, in order to increase the role of the private sector in the economy and society. Neoliberal ideology has undermined collective action to mitigate climate change, especially in the United States.[18] 'Hard' neoliberalists reject the redistribution of income except for a basic 'safety net'. They take the position that government exists only to maintain property rights, defend capitalists and maintain price stability. This ideology and its associated policies, which nowadays are widespread in the world, make life harder for the poor and weaken environmental protection.

Neoliberalism opposes government intervention in the economy on the false grounds, among others, that it reduces economic efficiency. This claim, which has enormous political implications, is supposedly based on neoclassical economics theory, but the truth is that it has been refuted by means of neoclassical economics analysis (see Box 7.1).

Box 7.1 Does Government Intervention Necessarily Damage the Economy?

Neoclassical economists postulate an ideal, economically efficient state called Pareto optimal or Pareto efficient. It's an economic state where resources cannot be reallocated to make one individual better off without making at least one other individual worse off. Incidentally, Pareto efficiency does not require an equitable distribution of wealth—it is only concerned with economic efficiency. Under the following set of assumptions that bear no relation to reality, neoclassical economics holds that an economic system will be Pareto optimal and therefore, by implication, desirable:

- a set of markets in economic equilibrium equilibrium, i.e., with supply equal to demand
- no externalities
- all participants have complete information
- no participant has the power to influence the price at which it sells a product or service
- no increasing returns to scale.

Pareto efficiency is not just a harmless mathematical exercise, because the concept is used to argue against government intervention in the market, e.g. by laws, regulations, taxes or subsidies. The (invalid) argument is that any government intervention takes the economic system further away from the ideal (for economists who disregard social equity), Pareto-efficient state. This notion was refuted in 1956 by economists Lipsey and Lancaster with their elucidation of the general theory of the second best.[19] The second best applies when one or more of the Pareto optimal assumptions are invalid, that is, in the real world. In non-technical language, Lipsey and Lancaster showed that, in such an imperfect market, introducing one or more additional market 'distortion' (e.g. a carbon tax or energy efficiency regulation) may partially counteract the existing imperfections and lead to a more efficient outcome. Despite this classic research that has never been refuted, supporters of neoliberalism continue to claim incorrectly that governments must not intervene in the market, or they will make it less 'efficient'.

This is in addition to the obvious point that, unlike the private sector, governments do not have to make a profit, and so a government-owned business may produce less expensive goods and services than one conducted by the private sector.

Many detailed critiques of neoclassical economics and/or neoliberalism have been published, showing that proposals to reject them have sound logical bases.[20] Alternative approaches to sustainability—such as ecological economics (discussed in the next section), ecofeminism[21] and deep ecology[22]—have different ethical bases from neoclassical economics.

Neoliberalism has failed in practice as well as in theory, as witness its inability to respond to the Global Financial Crisis of 2008–2009 and the COVID pandemic of 2020 onwards. To drive the economic recovery from both crises, governments that pay lip service to neoclassical economic theory and neoliberalism practice made huge expenditures using quantitative easing (defined below), violating their 'principles'. Despite or perhaps because of these failures, the proponents are defending their ideologies, although some members of the public are challenging the conventional claims (see the exchange in Box 7.2).

Box 7.2 A Defence of Neoliberalism and a Refutation

Neoclassical economist Richard Holden writes that neoliberalism, as he sees it, is *"Not the fanatical laissez-faire views that oppose government and market regulation. But a view of liberalism—in the classical sense, emphasising individual liberty—that harnesses the power of markets for social and economic good… But it also means a commitment to the hard work of building a more prosperous society, not just shrieking from the sidelines, complaining about the things that are wrong, and misidentifying the solution."*[23]

However, an online comment on the article by Tiffany Whistle rejects Holden's argument fiercely: *"Oh dear oh dear, what a privileged, entitled article. Pure ruling class propaganda, as you would expect here. Just work harder people, instead of 'shrieking' from the sidelines. Leave it all to the experts, and everything will be fine. What magical powers he ascribes to 'the market' rather than for instance, democratic reforms, the labour movement, mass protests, people getting arrested, imprisoned, even shot for standing up to government tyranny, you know, the things that actually advanced the interests of the poor. And I suppose he just forgot to tell you that the 'free market' is actually run by rich investors and used to further their own interests, at your expense. But don't complain, suck it up. Plus you get the inevitable false dichotomy, it's between liberalism and socialism, one good one bad."*

Based on the scholarly critiques of neoclassical economics and its neoliberalism offshoot, we believe that these ideologies are swindles that have been imposed on the people of the world and their governments by the rich and powerful. They provide a cunningly devised story to convince people that the socioeconomic inequities in society are natural and unavoidable, and that following current economic theory gives the best possible outcome for everyone.

Some of these critiques, and many others from Karl Marx onwards, go further to critique the system in which neoclassical economics and neoliberalism are embedded, namely capitalism. This is the economic and political system in which a country's trade and industry are controlled by private owners for profit, rather than by the state or workers. These profits are produced by exploitation of working people and the environment. The social conflicts this system creates are pacified by the promise of rising levels of material prosperity through economic growth.[24]

While we agree with many of the criticisms of capitalism, we are conscious of the imminent threats to civilisation and the need for urgent action to avoid its collapse and to transition as smoothly as possible to the Sustainable Civilisation. At present, an uprising against capitalism looks very unlikely, although it makes sense to prepare for its unexpected collapse resulting from crises of its own making, or war or pandemic. It also makes sense to design for the possibility of a smooth transition to a post-capitalist society.[25]

The approach of this book is to focus on transitioning away from neoclassical economics and neoliberalism, because we believe they are poorly based and realistically could be replaced within a generation, unlike capitalism. Nevertheless, widespread adoption of the strategies we propose, especially physical degrowth to a steady-state economy as discussed in the next sections, would weaken capitalism as well.

An alternative to neoclassical economics, with the goal of ecologically sustainable and socially just development, is the interdisciplinary or transdisciplinary field of ecological economics.

7.2 Ecological Economics and the Steady-State Economy

In the 1970s, when some neoclassical economists were developing environmental economics, an economist with a broader outlook, Herman Daly, was building on earlier writings to propose a more radical field of knowledge that he called 'steady-state economics' or SSE.[26] With inputs from other economists, such as Kenneth Boulding and Robert Ayres, and from scientists, it evolved into 'ecological economics', the economics of sustainability.

What Is Ecological Economics?

Unlike environmental economics, ecological economics is *not* a branch of neoclassical economics. It draws upon several disciplines and comprises several different strands. An important influence on the development of ecological economics was the work on 'ecoscience' by biologists Paul and Anne Ehrlich and physicist co-author John Holdren.[27] Among other things, they showed that, at one conceptual level, environmental impact could be expressed as population multiplied by GDP per person (affluence) multiplied by technological impact (environmental impact per unit of GDP), the well-known $I = PAT$ identity discussed in Sect. 2.1. Robert Costanza, the lead author of an important introduction to ecological economics, was originally an architect and then a systems ecologist. His book, co-authored by several economists, integrates ecology, economy and social justice.[28] It was followed by, among others, the modern textbook of ecological economics by Herman Daly and Joshua Farley—its approach is to modify neoclassical economics by limiting throughput and scale of the economy and ensuring just distribution of income and wealth.[29] Meanwhile, economist Manfred Max-Neef was working on revitalising low-income communities through 'barefoot economics', an approach with a strong social justice aspect.[30] These and many other contributors to the development of ecological economics broadened their own knowledge, and that of their readers, far beyond the disciplines of their original degrees. Nevertheless, the international journal *Ecological Economics* still seems to be dominated by economists, albeit mostly enlightened ones. Ecological economics is still under-represented in political economy,[31] and political science.[32] Multiple pathways are needed to reach the summit of Mount Sustainability.

Ecological economics recognises that "*the purpose of the economy is to maximise collective wellbeing, ensuring everybody has a chance to thrive... the human-constructed economy can and should serve humanity rather than the other way around*".[33] Implicit in this understanding is that the economy should enable the protection of the natural environment upon which humanity depends for its survival. An ecocentric approach would add that the environment and its living inhabitants have a right to exist apart from the services they give to us.[34] This can be contrasted with the neoclassical view that, for example, a forest has no value unless people are willing to pay to walk in it or until it is cut down for timber.

The essence of ecological economics is illustrated in Fig. 7.1, where the two systems of economics and environment of the neoclassical economics framework have been united under the single global biosphere. Economic activity and its impact on the environment can expand, but only within the physical boundary of the biosphere. Endless economic growth is ruled out. People are both consumers and citizens and so can cooperate as well as compete; they can respond to ethical principles as well as economic self-interest. Households as citizens value the environment in non-monetary terms, recognising that it has intrinsic value, an ethical position. And even a little scientific education or common sense should give us the understanding that the biosphere is our life support system—without it, we could not exist (Sect. 3.1).

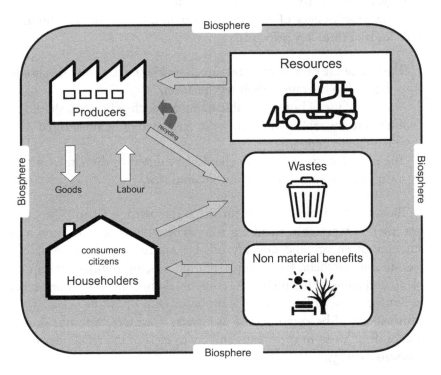

Fig. 7.1 The essence of ecological economics

While neoclassical economics mainly focuses on the efficient allocation of resources, ecological economics is concerned primarily with the wellbeing of the planet and especially its human inhabitants.[35] Therefore, its goals are to limit the scale of human activity to a magnitude that's ecologically sustainable and to ensure the distribution of resources is socially just. These goals come before efficient allocation of resources. They are in accord with the priorities of ecologically sustainability development discussed in Sect. 3.1 and illustrated in Fig. 3.1.

Limiting scale is one of the main differences between ecological and neoclassical economics. As current economic activity is already exceeding ecological limits (see Chap. 2), this entails ending global growth in the use of energy, materials and land, and in population. To allow for the resource needs of the poor countries, the rich countries and those rapidly becoming rich must follow programs of planned physical degrowth to ecologically sustainable socioeconomic systems.

Some practitioners of ecological economics[36] suggest that limiting scale leads to three key principles:

1. The rate of use of renewable resources, such as forests and fish, should be limited to the rate of their regeneration.
2. The rate of use of non-renewable resources, such as lithium, should be limited to the rate at which they can reclaimed or be substituted with renewable resources.
3. The rate of waste generation should be limited to the rate at which wastes can be assimilated by the environment.

The second principle is problematic, as discussed in Chap. 5, because not all non-renewable resources have renewable substitutes and vast amounts of energy and possibly other resources would be needed to reclaim some of them. For example, phosphorus, an essential nutrient for plants and animals, has no known renewable substitute. After human use in fertilisers and the diet, most phosphorus ends up in the oceans from which it would be very difficult to extract. Therefore, we must greatly reduce these losses to the oceans and try to reclaim the remainder using renewable energy.

Indicators of Wellbeing

Another important difference between ecological and neoclassical economics is in the way of measuring human wellbeing. Supporters of neoclassical economics place great emphasis on consumption as measured by Gross Domestic Product (GDP) or GDP per person, although they were never intended to be measures of wellbeing. Human wellbeing goes far beyond consumption measured in monetary terms, encompassing sufficient nutritious food, shelter, health, education, personal security, friendships, social support and basic energy supply. GDP arose *"after the carnage of the Great Depression and World War II"*.[37] One of its definitions is *"the total monetary or market value of all the finished goods and services produced within a country's borders in a specific period, normally one year"*. GDP counts all economic activity, whether it results from the clean-up of an environmental disaster, or the increase in hospital and funeral services following a war or pandemic, or the growth in the renewable energy industry. It is the one statistic to hold them all.

While GDP is a useful measure of total economic activity, it tells us very little about wellbeing which depends on the distribution of wealth and economic activity within the population, the population number, the diversity of industries and businesses, the natural resources available, the level of pollution of the environment, the health and education of the population, the availability of public facilities (such as public transport, education, housing, libraries, parks), the availability of technologies and other human-made capital, and the institutions of law and governance.

Ecological economics recognises alternatives to GDP. Economists who practice ecological economics want a more realistic economic indicator in monetary terms. The Genuine Progress Indicator (GPI) and the Index of Sustainable Economic Welfare (ISEW) are two of the preferred options.[38] GPI captures more costs and benefits than GDP and subtracts the costs from the benefits to provide a net outcome.[39] For rich countries, GPI and ISEW have tended to grow as GDP grows, but the growth in GPI and ISEW generally falls behind that of GDP and, in some countries, ceases. In Australia, the economic elements of GPI grew from 1962 to 2013, while the environmental elements declined and the social elements remained stable.[40] Another economic indicator is the Gini index, which measures the extent to which the distribution of income among individuals or households within an economy deviates from a perfectly equal distribution. A Gini index of 0% represents perfect equality, while an index of 100% represents perfect inequality. For example, South Africa had a Gini index of 63% in 2014, the United States had 41% in 2018 and Finland 27% in 2018.[41]

Scientists and other non-economist practitioners of ecological economics want measures of wellbeing that are not limited to monetary wealth and income—for them a single indicator will not suffice. The OECD recognises numerous dimensions to wellbeing, in addition to income and wealth, including work and job quality, housing, health, environmental quality, safety and social connections.[42]

The main differences between ecological economics and neoclassical economics are summarised in Table 7.1.

Ethical Differences

Substantial differences in values and ethics are implicit in the differences set out in Table 7.1. These differences are in the underlying philosophies, the goals of economics, the value and role of the natural environment, the value and role of people, and how human wellbeing should be measured. Neoclassical economics is based on the philosophy of utilitarianism, the belief that decisions over resources should be made by comparing the utilities of the people affected by those decisions. This is related to instrumentalism: the belief that the environment has no intrinsic value; that it only has value to the extent that it is valuable to humans. This is also an anthropocentric (human-centred) belief. Despite the value judgements in the above beliefs, the proponents of neoclassical economics persist in claiming their discipline is purely objective or value free.

Ecological economics too is based on subjective or normative assumptions or beliefs, ones that we authors support, such as intergenerational and social equity, and its conceptions of the relationship between humans and the environment. It is less anthropocentric than neoclassical economics.

Table 7.1 Comparison of the main features of neoclassical and ecological economics

Neoclassical economics	Ecological economics
Neoclassical economics is a discipline; one of its sub-disciplines is environmental economics	Ecological economics is an interdisciplinary or transdisciplinary field
Principal goal is the efficient allocation of limited resources	Principal goals are the wellbeing of the planet and its human inhabitants
It assumes that its goal is achieved automatically by people maximising their respective utilities (i.e. utilitarianism) and businesses maximising profits; but in practice this maximises wealth and income for a tiny minority	It pursues its goals by proposing to limit the scale of human activity to a magnitude that's ecologically sustainable and to ensure the distribution of resources is socially just
Instrumentalism: The belief that the environment has no intrinsic value; it only has value to the extent that it is valuable to humans	Some strands of ecological economics accept the environment has intrinsic value
Environment is an infinite resource base, waste sink and source of amenities	Environment is the life support system of the planet and therefore of civilisation. Its capacity is finite.
Unlimited economic growth is possible and desirable	The economy is bounded by finite capacity of the planet to provide resources and assimilate wastes; a steady-state economy with low throughput is necessary for sustainability
People and households are consumers who compete to maximise their utility; businesses to maximise profits	People and households are both consumers and citizens; they compete and cooperate; businesses have social responsibility as well as responsibility to shareholders
The key indicator of economic performance is GDP; in practice it is also used as an indicator of wellbeing	Better economic indicators include GPI, ISEW and Gini. Additional, non-economic indicators are needed to measure wellbeing.
It lacks a theory of money creation	Its practitioners have various theories of money creation

Note: Among the several different approaches to ecological economics, we only consider here those that accept the limits to growth, and the need for a steady-state economy and social justice. These are essential for 'strong' sustainability

Ethics is concerned with conduct that is right or wrong, good or bad. Since the proponents of each of these approaches believe their way is right, they are taking ethical positions. There is a growing literature on ecological ethics.[43]

7.3   Planned Degrowth[44]

Planetary boundaries are being exceeded (see Chap. 2). At one conceptual level, the principal drivers of these overshoots are growth in consumption per person and population, together with polluting technologies. This growth must be reversed and the global socio-economy as a whole must undergo physical degrowth to a SSE. If we fail to do this, then nature will do it for us through a series of disasters leading to the collapse of civilisation. Indeed, nature's reaction has already begun with droughts, firestorms and floods of increasing frequency and severity.

Unfortunately, the term 'degrowth' triggers negative reactions in people who have been conditioned to believe that economic growth is necessary for increased wellbeing. Herman Daly used the term 'qualitative growth', that is, growth in human wellbeing, as opposed to quantitative growth as measured by GDP. In the interests of brevity and saying what we mean, we will use 'planned degrowth', while distinguishing between physical and monetary degrowth (see below).

Planned degrowth is *not* the same as economic collapse, which can result from an ecological disaster,[45] the failure of neoclassical economics in a financial crisis, war, internal conflict, and poor governance, including economic and political exploitation.[46] Planned degrowth, as conceived by many ecological economics practitioners, is *a planned program to reduce the use of energy, materials and land, and to stabilise population, initially in the rich countries.* The goal of planned degrowth is a sustainable, steady-state economy. To be truly sustainable, planned degrowth must increase human wellbeing while protecting and restoring the environment, in other words, to foster sustainable prosperity. Social justice demands that low-income countries must grow their economies and hence consumption. Therefore, high-income countries must undergo planned degrowth. In the saying attributed to Mahatma Gandhi: "*The rich must live more simply so that the poor can simply live*". The following definition of planned degrowth by economic anthropologist Jason Hickel makes these considerations explicit: *"Degrowth is a planned reduction in less-necessary production in rich countries that is socially just and achieved in a democratic manner"*.[47]

We have defined planned degrowth in biophysical terms, because that's the environmental imperative. Of primary importance is to develop a strategy to protect our life-support system and facilitate the wellbeing for all people (see Sect. 7.5). The fate of GDP is of secondary importance. Within a sustainable degrowth program, some industries will expand—for example, renewable energy, energy efficiency, electric vehicles, public transport, bicycles, aged care, child care, public health facilities, housing, medical care, public education and training, the arts, nature conservation, and plantation forestry in appropriate locations—while others will contract—for example, fossil fuels, road-building, logging of native forests, high-GHG-emission agriculture, tobacco, armaments for attack rather than defence, advertising and financial services. The net effect must be to reduce biophysical impacts and this may in turn reduce GDP. In contrast, green growth would allow GDP and hence net biophysical impacts to increase, albeit not as rapidly as business-as-usual growth.

To what extent does physical degrowth imply monetary degrowth, that is, a reduction in GDP? In the jargon of economics, to what extent is there decoupling between monetary economic activity and physical economic activity and hence environmental impact? Several studies have established that *on average* GDP and biophysical impact are coupled, although there are exceptions observed over short periods of time in specific locations for particular environmental impacts.[48] Clearly, all economic activity depends on at least some physical activity and hence has some environmental impact. For example, education in a school requires a building and equipment for teaching and learning. Writing a book nowadays requires a computer, internet connection and a server. However, the total, life-cycle environmental impacts of school-teaching and book-writing are generally much less than those of owning and operating a private jet aircraft. Therefore, while shifting to 'greener' economic activities will result in less environmental impact (i.e. relative decoupling), there is no absolute decoupling. Therefore, green growth must be rejected as insufficient. The Sustainable Civilisation needs both greener economic activities and planned degrowth.

Environmental economist Peter Victor investigated scenarios for no-growth and monetary degrowth of the Canadian economy, defined in terms of a reduction in GDP. Creating a macroeconomic model, Victor found that, as expected, simply reducing GDP resulted in increasing unemployment. However, in scenarios where reducing GDP was combined with other policies—for example, working time reduction, increased government expenditure on poverty reduction and health care, government investment in green infrastructure, ending population growth—an economy with reduced GHG emissions, reduced unemployment and reduced poverty can be achieved.[49]

An alternative approach was taken by Graham Turner, using a bio-physical method, the Australian Stocks and Flows Framework, to model physical degrowth in the Australian socio-economy.[50] Stocks are quantities of physical items, such as land, livestock, people and buildings at a point in time. Flows are the rates of change over a period, for example, net addition of agricultural land, new computers, births, deaths and immigration. Although he used an entirely different method, Turner obtained similar results to Victor's.

Simone D'Alessandro and colleagues constructed a simulation model for planned degrowth in France that combined environmental and radical social justice policies, together with a reduction in consumption and exports. The environmental policies included greater energy efficiency, renewable energy and a carbon tax. The social justice policies comprised a job guarantee, working time reduction and a wealth tax. They obtained a viable sustainable economy at the 'cost' of substantial levels of public expenditure.[51] While cost may be a political problem for governments that follow the prescriptions of neoclassical economics and neoliberalism, it is not necessarily an economic problem within the macroeconomics framework of Modern Monetary Theory (MMT), as discussed in the next section.

7.4   Paying for the Transition: MMT

The myths of neoclassical macroeconomics are abundant in popular political and economic beliefs about money. Here is a string of them:

Like a household, a national government must balance the budget. Failure to do so imposes debt on future generations. The government cannot afford to spend more on health, education or housing without raising taxes or withdrawing funds from another important sector such as defence, or by borrowing. If a government has a large deficit or creates a lot of money, this will inevitably cause inflation. Or the government will go bankrupt.

Several years ago, we authors assumed uncritically that these statements may be true. But our knowledge of the absurd assumptions of neoclassical economics outlined in Sect. 7.1 stimulated us independently to re-examine conventional beliefs about money. We discovered a literature written by a minority of economists, some eminent, setting out a clear, logical account of macroeconomics according to Modern Monetary Theory (MMT).[52] Popular introductions have been given in videos by Stephanie Kelton, Alan Kohler and Steven Hail.[53]

MMT shows that, for countries with monetary sovereignty (see Box 7.3), all the above beliefs are wrong. National governments of these countries do not have to balance the budget and in practice rarely do so. Budget deficits are often necessary and beneficial: they are merely evidence of extra government spending, which boosts the wealth of private sector businesses and households. The outcome depends on what deficit spending is used for.

Box 7.3 Monetary Sovereignty

A fully monetary sovereign government:

- issues its own currency
- collects taxes in that currency
- maintains a floating exchange rate
- avoids borrowing in foreign currencies, and
- is not heavily dependent on imported necessities priced in foreign currencies.[54]

A floating exchange rate means that the government has no obligation to exchange its currency at a fixed rate for any foreign currency or commodity such as gold or anything in limited supply. Examples of monetary sovereign countries are the United States, United Kingdom, Japan, China and Australia. However, state or provincial governments and the member governments of the European Union are *not* monetary sovereign—they do have to balance income and expenditure, or to borrow money to cover a deficit.

According to MMT, governments do not have to borrow money that they themselves create. They can create debt-free money upon which there are no *financial* limits. However, there must be limits to ensure that government spending, more accurately, total spending by the government and private sectors, does not exceed the productive capacity of the economy. Failing to satisfy that constraint may lead to inflation. The threat of inflation is the main criticism of MMT from neoclassical economists. Let's examine this important issue more closely.

Inflation is a continuous increase in the cost of living as the prices of goods and services rise. It leads to a decline in the value of money; that is, a given amount of money cannot buy as much today as it did previously. Economists recognise two mechanisms of inflation, cost-push and demand-pull. *Cost-push inflation* is when an increase in prices is caused by an increase in the cost of production. *Demand-pull inflation*, the mechanism of interest for MMT, is when an increase in prices is caused by rising aggregate demand and firms pushing up prices due to the shortage of goods.[55]

A big increase in government spending—for example, to provide the infrastructure for transitioning from fossil fuels to renewable energy, or to provide public housing for the homeless—has the risk of causing demand-pull inflation if these programs exceed the productive capacity of the economy. For example, if the country does not have a sufficiently skilled labour force or access to the raw materials or the technologies for its programs, then the prices of these essential elements will rise. From the limited perspective of neoclassical economics, these programs can be funded in a non-inflationary way either by reducing funding for other government programs, or increasing taxation to pay for the new programs, or increasing interest rates to reduce borrowing and spending by the private sector, or borrowing the money from the private sector of the country[56] or from overseas.

In the interpretation of macroeconomics of MMT, the government asks the central bank to create debt-free money to fund the programs and then, if it becomes necessary, applies one or more of several ways of avoiding inflation from the increased spending. The most obvious way is to increase taxation. In the MMT perspective, taxation does not pay for the new programs, instead it is used to reduce aggregate demand and hence spending. If taxation were the only way of combatting inflation from the new programs, a sceptic could ask, "What's the point of MMT? Taxation still has to be increased and so it doesn't matter whether we believe that its purpose is to fund the programs or combat inflation." However, other ways of reining in demand-pull inflation can be used, such as wage and price controls, rationing, and incentives for saving.[57]

Even better is to use part of the new government spending to increase the productive capacity of the economy, for example, by subsiding the education and training of workers to learn the necessary skills and to manufacture of some of the technologies instead of importing them.

Before even considering these anti-inflationary measures, MMT proponents point out that we need a *"careful accounting of the resources that can be made available and to weigh those against what will be needed. Only then should we address the question of whether taxes and other means might be needed to reduce income and private spending sufficiently to avoid inflation"*[58] as the new programs are phased in. This was the approach of John Maynard Keynes in his study on how the British Government could pay for fighting World War II.[59]

The empirical evidence, that the inflationary risk of big government spending has been exaggerated, is given by the experience of combatting COVID-19 from 2020 onwards and the economic recovery from the Global Financial Crisis (GFC) of 2008–2009. While fighting the pandemic, many governments have each spent hundreds of billions of US dollars equivalent in their own currencies to maintain their economies. From March 2020 to March 2022, the US government spent trillions of US dollars, bringing its total deficit to US$9 trillion.[60] Yet demand-pull inflation has not eventuated. The only inflation observed has been cost-push inflation resulting from supply shortages during the economic recovery from the pandemic and the war in Ukraine.

The governments injected these huge amounts of money into the economy by a form of monetary policy known as *quantitative easing* (QE). Governments purchase government long-term bonds (see Box 7.4) or other financial assets from domestic financial institutions. In neoclassical economics theory, this injects money into the economy and lowers interest rates, thus stimulating the economy.[61] In terms of being potentially inflationary, spending money acquired by QE is no different from spending money created by a monetary sovereign government that understands MMT. The difference is that money spent under QE is a loan from the private sector to the government that must be paid back when the bonds mature, while the MMT approach does not involve a loan and so does not contribute to the government deficit and does not pay interest to the private sector. A monetary sovereign government does not need to issue bonds to fund spending.[62]

Box 7.4 Bonds in Economics

A bond is a unit of debt issued by a government or company. For the bond-holder, it is an investment that provides a return in the form of fixed periodic interest payments and the eventual return of the principal at maturity. It represents wealth for the bond-holder. For the issuer, it is considered by neoclassical economists to be a loan it receives from the bond-holder. The issuing government or company can use the borrowed funds to finance projects and operations, although a monetary sovereign government does not need to borrow its own currency.

The above standard account of bonds is correct when the bond issuer is a non-monetary sovereign government or corporation. However, an MMT perspective has a different explanation of the process when a monetary sovereign government issues bonds in its own currency. Then bond issuance does not involve a net issuance of debt. It simply exchanges reserves at the central bank for treasury bonds. These are both financial liabilities of the government sector, so they are a swap of one form of government sector liability for another. Currency is a government IOU. You cannot meaningfully borrow your own IOU.

According to MMT, the issuance of bonds by monetary sovereigns can play three roles. Firstly, it can drain reserves, to assist the central bank in hitting its inflation target. Secondly, it offers a safe financial asset to fund managers, with generally a higher interest rate than bank deposits. Thirdly, bond programs define the term structure of default risk-free interest rates.[63]

Additional empirical evidence that government deficits are not necessarily highly inflationary comes from the budgetary trends. For many developed countries, deficits have been the normal situation for decades without driving a high rate of inflation.[64]

Some critics of MMT claim the creating debt-free money will lead to hyperinflation as occurred in Zimbabwe in the first decade of the twenty-first century. However, this hyperinflation was "caused mostly by social unrest, the collapse of agriculture and hence major supply shortages, and consequently, heavy external debt".[65] The government did not have monetary sovereignty as defined above.

Clearly, the standard objection to the MMT account of macroeconomics, that it is necessarily inflationary, fails in the face of empirical evidence as well as theory. That's the scientific approach. Even leaders of the economy have occasionally admitted that the orthodox approach is incorrect. For example, the then chairman of the New York Federal Reserve, J. Beardsley Ruml, stated in 1946 that *"a sovereign national government is finally free of money worries and need no longer levy taxes for the purpose of providing itself with revenue … It follows that our Federal Government has final freedom from the money market in meeting its financial requirements."*[66] The eminent US economist, Paul Samuelson compared the myth that the US federal government could run out of its own currency to an old-fashioned religion.[67] For a monetary sovereign government, money is never a scarce resource. It is real resources and their current productive potential that are limited in supply.[68]

This leads to the question: why don't the governments of monetary sovereign economies acknowledge MMT and the opportunity it offers to fund environmental protection, public education, health and housing properly? One answer is that the state has been captured by the neoclassical economics ideology and its associated political power structure. The small minority of politicians, their economic advisors and public servants who understand that the conventional dogma is incorrect do not dare to risk public ridicule or damage to their careers by supporting MMT. However, it appears from the money creation programs being implemented during the economic recovery from the pandemic, that huge budgetary deficits are no longer regarded as a major problem by some governments.[69] *This suggests that MMT is being adopted by stealth.*

Another answer has been given by critics of MMT: if governments accept publicly that there are no financial limits on spending, they could spend excessively and irresponsibly, without considering the productive capacity of the economy. Paul Samuelson actually said that it's better to maintain the myth that the national budget must be balanced by scaring politicians into behaving responsibly, like "old-fashioned religion".[70]

MMT supporters reply that government spending could be overseen by an independent commission or an existing institution with enhanced powers such as the Congressional Budget Office in the USA, the Office of Budget Responsibility in the UK and the Parliamentary Budget Office in Australia.

A basic understanding of MMT suggests how we could fund an important potential policy that combines ecological sustainability with social justice, namely a *job guarantee* (JG). It's a permanent scheme that offers the opportunity for socially productive work to otherwise unemployed people who wish to work full-time or part-time. It would be funded by the national government, using money creation according to MMT principles and would pay the basic minimum wage. The jobs would be offered and administered by all levels of government and by registered non-profit community NGOs. The work would be socially productive, but not profitable to private businesses, so the scheme would not compete with the market. Jobs would offer basic training and could include, for example: revegetation of public land; maintenance of national parks; assisting scientists to monitor air and water pollution; staffing food banks, recycling depots, women's refuges and refuges for homeless people; providing free meals for people in need; supporting fire-fighters and other emergency response operations; and funding citizen scientists to assist in university research projects in epidemiology, public health, environmental protection and social equity.

The JG would also stabilise the economy by employing more people during downturns, maintaining or enhancing skills until the economy recovers, and employing fewer people during boom times when better-paid jobs become available in the private sector. Thus government expenditure on the JG increases automatically when it is needed (when the number of job seekers increases) and so does not drive inflation at that time; it decreases when it is not needed (when workers leave the program) and so does not drive inflation at that time either. There can be no inflationary pressures arising directly from a policy where the government offers a fixed wage to any labour that is unwanted by other employers. The JG is preferable to the situation under neoclassical economics where stabilisation of the economic cycle is left to the involuntary unemployed.

The JG has several advantages over a universal basic income (UBI). As well as its role as an effective buffer for the economic cycle, the JG offers socially productive work to those who want it. Because it is targeted at the involuntary unemployed instead of the whole population, it can afford to pay a decent basic wage, unlike the UBI. It increases the productive capacity of the economy.

7.5 Strategy for Planned Degrowth to Sustainable Prosperity

The basic strategy foreshadowed in Sects. 7.2, 7.3, and 7.4 is eight-fold. First, we must discard the notion that GDP is relevant to human wellbeing and replace it with a set of direct indicators of different aspects of wellbeing, such as health, education, housing, aged care, child care, personal security, democratic participation, air quality, water quality, national park area per person, unemployment rate, and so on. More generally, neoclassical economics must be replaced with ecological economics, and ecological economics should be strengthened with the insights of MMT.

Second, we need policies to encourage businesses and industries to provide environmentally sound, healthy products, and to discourage pollution and waste of natural resources. Such policies include: environmental taxes (including a carbon tax)[71]; environmental audits and labelling of goods and services; government grants, loans, fees, rebates and subsidies; government funding for research, education, training and information; regulations and standards for energy efficiency and materials efficiency; legislation to encourage design of products for reuse, recycling and remanufacture; and tighter controls on advertising harmful products.

Third, the policies for the substitution of green economic activities for polluting and wasteful activities must be implemented within a framework of transitioning to a steady-state economy. For the high-income countries, this entails planned degrowth, at least in biophysical terms. National and international caps must be placed on the throughput of materials and energy and a fair distribution of these limited resources must be implemented. This would be resisted strongly by big business and the governments they have captured. Nevertheless, as we are facing a climate emergency, a start can and must be made by reducing total global energy consumption so that renewable energy can rapidly replace all fossil fuels, as discussed in Sect. 4.5. Initially, the high-income countries could lead the way. Between 2000 and 2019, the European Union has been reducing its total final energy consumption and there has only been a slight increase in the total for the OECD over 2004–2019. However, this is not so impressive when we consider that much energy is embodied in goods manufactured in rapidly growing economies and imported into the European Union and OECD countries. The rich countries must make much more substantial reductions in their respective consumptions of energy, for example, halving them by 2050. To make matters worse, recalcitrant rich countries—such as the USA, Canada and Australia –are actually increasing their respective territorial energy consumptions.[72] They can be pressured to cut excess consumption by a Carbon Border Adjustment Mechanism.[73] The environmental impacts of the rich could also be limited by environmental taxes, a wealth tax (e.g. in the form of an annual property tax), an inheritance tax and a tax on business-class air travel.

Fourth, national and subnational policies are needed to retrain, relocate or provide retirement pensions for workers who lose jobs as a result of the transition. This is particularly important for workers in declining industries such as fossil fuels and in industries that should decline such as native forest logging.

Fifth, the national government must fund a job guarantee at the basic wage for all adults who want to work but cannot find a job in the formal economy (Sect. 7.3).

Sixth, so that high incomes are not essential for the wellbeing of citizens, governments should expand public goods and services: health, education, housing, transport, town parks, national parks and libraries. More generally, governments should provide universal basic services (UBS) in which a social wage replaces part of a monetary wage.[74] UBS includes the public goods and services listed above.

Seventh, because degrowth is likely to reduce the available work in the formal economy, a shorter working week should be implemented to share the work around. The job guarantee will also assist.

Eighth, non-coercive policies should be implemented to end population growth, especially in the rich countries where per capita consumption is very high. Policies needed in the rich countries include improved social security, removal of tax incentives for births, and public education. In the poorer countries, the well-known incentives are social security, contraceptive education and availability, and education of women.

The numbering of the above items is for identification and is not intended to indicate priority or order of implementation. Ideally, they would be implemented simultaneously.

Several of the items are mutually reinforcing in fostering degrowth. Because energy consumption is closely coupled to economic activity (see Chap. 4), reducing the former will reduce the latter and hence biophysical impacts. Similarly, stabilising population will remove one of the drivers of economic and hence biophysical growth. Assisting workers who lose jobs as a result of the transition will reduce public resistance to the transition. Shifting attention away from GDP to genuine indicators of wellbeing will increase public support for the transition as improvements in their quality of life become visible. UBS would have increased impact if linked to a job guarantee.[75] In countries with monetary sovereignty, funding a job guarantee and UBS by creating money according to Modern Monetary Theory will also increase public support, provided misleading hostile propaganda can be overcome.

7.6   Conclusion

The Sustainable Civilisation needs an economic system that fosters ecological sustainability and social justice. The current dominant system, neoclassical economic theory together neoliberalism practice, is based on numerous myths. Its practitioners claim it's a science although it does not stand up the scientific scrutiny. The interdisciplinary field of ecological economics offers a better alternative conceptual framework. It recognises that endless growth on a finite planet can only lead to disaster and therefore its priorities are to limit the scale of human activity to a magnitude that's ecologically sustainable and to ensure the distribution of resources is socially just. The efficient allocation of resources comes a distant third.

From the viewpoint of ecological economics, the important question for the rich countries is how to undertake planned degrowth to a steady-state economy while achieving wellbeing for all.

Scenario modelling by several different methods finds that simply cutting consumption and doing nothing else leads to unemployment and poverty, as expected. However, the modelling also finds that an ecologically sustainable society that is more socially just can be reached by implementing a mix of different government policies to simultaneously reduce consumption and greatly improve social justice, thus reducing the perceived need for high incomes and hence high consumption. The emphasis must be on 'sustainable prosperity', that is, human and planetary wellbeing. We must discard the notion that GDP is relevant to human wellbeing and replace it with a set of direct indicators of different aspects of wellbeing.

To reduce consumption, policies could include caps on global resource use; taxation of pollution, wealth and inheritance; environmental audits and labelling of goods and services; legislation to encourage design of products for reuse, recycling and remanufacturing; and regulations and standards for energy efficiency and materials efficiency.

Social justice can be strengthened while reducing consumption by assisting workers who are disadvantaged by the transition, creating a job guarantee, reducing the working week, stabilising population size, and implementing universal basic services.

The transition can be funded by discarding myth-based neoclassical macroeconomics and replacing it with the insightful Modern Monetary Theory. This allows monetary sovereign governments to create money instead of borrowing it, subject to remaining within the productive capacity of the economy. Part of the created money can be used for investments in infrastructure and people that expand the productive capacity of the economy. The responsibility falls on monetary sovereign governments to assist low-income countries that are not monetary sovereign.

This strategy proposed for planned degrowth to a steady-state economy is compatible with a market economy that is more strongly constrained by national governments and international agreements than the present system. It is a feasible way forward. The most difficult part is overcoming the myths and pseudo-science supporting the existing economic system.

Notes

1. Steve Keen (2011). *Debunking economics: the naked emperor dethroned?* 2nd ed., Zed Books.
2. Richard Denniss (2022). Scott Morrison's economic lies. *The Saturday Paper*, 19 March, https://www.thesaturdaypaper.com.au/opinion/topic/2022/03/19/scott-morrisons-economic-lies/164760840013539; Richard Dennis (2021). *Econobabble: How to decode political spin and economic nonsense.* Black Inc.
3. Richard Denniss (2022), *op. cit.*; Richard Dennis (2021), *op.cit.*
4. Geoff Davies (2004). *Economia: New economic systems to empower people and support the living world.* ABC Books.
5. Larry Beinhart (2020). What is the Nobel Prize for economics really about? *Aljazeera*, 22 October. https://www.aljazeera.com/opinions/2020/10/22/what-is-the-nobel-prize-for-economics-really-about
6. John Blatt (1983). *Dynamic Economic Systems: A post-Keynesian approach.* M.E. Sharpe & Wheatsheaf Books.
7. Mark Diesendorf (1998). Australian economic models of greenhouse abatement. *Environmental Science & Policy* 1:1–12.
8. For example, William Nordhaus (1991). To slow or not to slow: the economics of the greenhouse effect. *The Economic Journal*, 101:920–937.
9. Steve Keen (2021). The appallingly bad neoclassical economics of climate change. *Globalizations* 18:1149–1177. https://doi.org/10.1080/14747731.2020.1807856p; Steve Keen (2022). Economic failures of the IPCC process. In: Stephen Williams & Rod Taylor (eds) *Sustainability and the New Economics: Synthesising Ecological Economics and Modern Monetary Theory.* Springer, chapter 10.
10. Michael Jacobs (1991) *The Green Economy: Environment, sustainable development and the politics of the future.* Pluto Press.
11. Mark Diesendorf (2014). *Sustainable Energy Solutions for Climate Change.* UNSW Press & Routledge, Chapters 4 & 8.
12. Hugh Saddler (2021). *National Energy Emissions Audit Report.* The Australia Institute, Figure 3. https://australiainstitute.org.au/report/national-energy-emissions-audit-september-2021
13. Europa (2022). Review of the EU ETS. https://www.europarl.europa.eu/RegData/etudes/BRIE/2022/698890/EPRS_BRI(2022)698890_EN.pdf

14. Clean Energy Wire (2021). Understanding the European Union's Emissions Trading Scheme. https://www.cleanenergywire.org/factsheets/understanding-european-unions-emissions-trading-system

15. Lucas Chancel & Thomas Piketty (2015). Carbon and inequality: from Kyoto to Paris. Paris School of Economics. https://doi.org/10.13140/RG.2.1.3536.0082; Helmut Haberl (2020). A systematic review of the evidence on decoupling of GDP, resource use and GHG emissions, part II: synthesizing the insights. *Environ. Res. Lett.* 15:065003. https://doi.org/10.1088/1748-9326/ab842a; Thomas Wiedmann et al. (2020). Scientists' warning on affluence. *Nature Communications* 11:3107. https://doi.org/10.1038/s41467-020-16941-y

16. Raphael Zeder (2020). https://quickonomics.com/do-we-really-need-economic-growth

17. John Quiggin (2022). The evolution of neoliberalism. In: Stephen Williams & Rod Taylor (eds) *Sustainability and the New Economics: Synthesising Ecological Economics and Modern Monetary Theory.* Springer, Chapter 6.

18. Anders Fremstad & Mark Paul (2022). Neoliberalism and climate change: How the free-market myth has prevented climate action. *Ecological Economics* 197:107353. https://doi.org/10.1016/j.ecolecon.2022.107353

19. R.G. Lipsey & Kelvin Lancaster (1956). The general theory of the second best. *Review of Economic Studies.* 24(1):11–32. https://www.jstor.org/stable/2296233

20. For example, Marilyn Waring (1988). *Counting for Nothing: What men value and what women are* worth. Allen & Unwin; Herman Daly & John Cobb (1990). *For the Common Good: Redirecting the economy toward community, the environment, and a sustainable future.* Green Print; Clive Hamilton (1997) *op. cit.*; Geoff Davies (2004), *op. cit.*; Sharon Beder (2006). *Free Market Missionaries: The corporate manipulation of community values.* Earthscan; Lee Boldeman (2007). *The Cult of the Market: Economic fundamentalism and its discontents.* Canberra: ANU Press; Naomi Klein (2008). *The Shock Doctrine: The rise of disaster capitalism.* Penguin; Steve Keen (2011). *Debunking economics: the naked emperor dethroned?* 2nd ed., Zed Books; Kerryn Higgs (2014). *Collision Course: Endless growth on a finite planet.* MIT Press; Richard Denniss (2021) *op. cit.*

21. Maria Mies & Vandana Shiva (1993). *Ecofeminism*. Zed Books; Iris Derzelle (2021). *Françoise d'Eaubonne's Ecofeminism: An overlooked left wing perspective*. https://booksandideas.net/Francoise-d-Eaubonne-s-Ecofeminism.html

22. Arne Naess (1988). Deep ecology and ultimate premises. *The Ecologist* 18(4–5):128–131; Bill Devall & George Sessions (1985) *Deep Ecology: Living as if nature mattered*. Gibbs M. Smith.

23. Richard Holden (2022). Vital signs: what the neoliberalism-hating left should love about markets. *The Conversation*, 11 March. https://theconversation.com/vital-signs-what-the-neoliberalism-hating-left-should-love-about-markets-178777 plus an online response.

24. Matthias Schmelzer et al. (2022). *The Future is Degrowth: A guide to a world beyond capitalism*. Verso.

25. Samuel Alexander (2020). Post-capitalism by design not disaster. *The Ecological Citizen*, 3(Supple B):13–21; Samuel Alexander et al. (eds) (2022). *Post-Capitalist Futures: Paradigm, politic, and prospects*. Palgrave Macmillan.

26. Herman Daly (1997). *Steady-State Economics: The economics of biophysical equilibrium and moral growth*. WH Freeman & Company.

27. Paul Ehrlich, John Holdren & Anne Ehrlich (1978). *Ecoscience: Population, resources, environment*. WH Freeman & Company.

28. Robert Costanza et al. (2014). *An Introduction to Ecological Economics*. 2nd ed. Taylor & Francis.

29. Herman Daly & Joshua Farley (2011). *Ecological Economics: Principles and applications*. 2nd ed., Island Press.

30. Manfred Max-Neef (1992). *From the Outside Looking In*. Zed.

31. An exception is Paul Burkett (2006). *Marxism and Ecological Economics: Toward a red and green political economy*. Brill.

32. An exception is Susan Baker et al. (eds) (1997). *The Politics of Sustainable Development: Theory, policy and practice within the European Union*. Routledge.

33. Philip Lawn & Stephen Williams (2022). An introduction to ecological economics: principles, indicators, and policy. In: Stephen Williams & Rod Taylor (eds) *Sustainability and the New Economics: Synthesising Ecological Economics and Modern Monetary Theory*. Springer, Chapter 11.

34. Eleanor Ainge Roy (2017). *The Guardian* Australia, 16 March. https://www.theguardian.com/world/2017/mar/16/new-zealand-river-granted-same-legal-rights-as-human-being

35. Hence ecological economics has been criticised for being anthropocentric, rather than ecocentric. However, as human wellbeing depends on planetary wellbeing, protecting the natural environment is an essential part of ecological economics.

36. Philip Lawn & Stephen Williams (2022), *op. cit.*

37. Elizabeth Dickinson (2011). GDP: a brief history. *Foreign Policy* https://foreignpolicy.com/2011/01/03/gdp-a-brief-history

38. Philip Lawn (2003). A theoretical foundation to support the Index of Sustainable Economic Welfare (ISEW), Genuine Progress Indicator (GPI), and other related indexes. *Ecological Economics* 44:105–118; Ida Kubiszewski et al. (2015). Estimate of the Genuine Progress Indicator (GPI) for Oregon from 1960–2010 and recommendations for a comprehensive shareholder's report. *Ecological Economics* 119:1–7.

39. Philip Lawn & Stephen Williams (2022), *op. cit.*

40. Daniel Kenny et al. (2019). Australia's genuine progress indicator revisited (19622013). *Ecological Economics* 158:110.

41. World Population Review. https://worldpopulationreview.com/country-rankings/gini-coefficient-by-country

42. OECD. https://www.oecd.org/wise/measuring-well-being-and-progress.htm

43. Patrick Curry (2011). *Ecological Ethics: An introduction.* 2nd edition, Polity Press; Holmes Rolston III (2012). *A New Environmental Ethics: The next millennium for life on Earth.* Routledge; Haydn Washington (2018). *A Sense of Wonder towards Nature: Healing the planet through belonging.* Routledge, Chapter 4.

44. Giorgos Kallis et al. (2020). *The Case for Degrowth.* Wiley; Matthias Schmelzer et al., *op. cit.*

45. Jared Diamond (2005). *Collapse: How societies choose to fail or survive.* Allen Lane.

46. Daron Acemoglu & James A. Robinson (2012). *Why Nations Fail: The Origins of Power, Prosperity, and Poverty.* Crown Business; for a summary see: 10 reasons why countries fall apart. *Foreign Policy*, 1 July 2012.

47. Paraphrased from the following debate between Jason Hickel and Sam Frankhauser: https://www.youtube.com/watch?v=YxJrBR0lg6s

48. Corinne Le Quéré et al. (2019). Drivers of declining CO_2 emissions in 18 developed economies. *Nature Climate Change* 9:213–217; Timothée Parrique et al. (2019). Decoupling debunked: Evidence and arguments

against green growth as a sole strategy for sustainability. European Environmental Bureau. https://eeb.org/decoupling-debunked1/; Helmut Haberl et al. (2020). A systematic review of the evidence on decoupling of GDP, resource use and GHG emissions, part II: synthesizing the insights. *Environ. Res. Lett.* 15:065003. https://doi.org/10.1088/1748-9326/ab842a

49. Peter Victor (2012). Growth, degrowth and climate change: A scenario analysis. *Ecological Economics,* 84:206–212; Peter Victor (2019). *Managing Without Growth, Second Edition: Slower by Design, Not Disaster.* Edward Elgar.

50. Graham Turner (2011). Consumption and the environment: impacts from a system perspective. In: Peter Newton (ed.), *Urban Consumption.* Collingwood: CSIRO Publishing, chapter 4. Reprinted as Turner (2016). Physical pathway to a steady state economy. In: Haydn Washington & Paul Twomey (eds) *A Future beyond Growth: Towards a steady state economy.* Earthscan from Routledge, pp.112–128.

51. Simone D'Alessandro et al. (2020). Feasible alternatives to green growth. *Nature Sustainability* 3:329–335. https://doi.org/10.1038/s41893-020-0484-y

52. Stephanie Kelton (2020). *The Deficit Myth: Modern Monetary Theory and the Birth of the People's Economy.* Public Affairs; William Mitchell, L. Randall Wray & Martin Watts (2019). *Macroeconomics.* Macmillan International & Red Globe Press; Steven Hail (2022). Paying for a Green New Deal: An introduction to Modern Monetary Theory. In: Stephen Williams & Rod Taylor (eds) *Sustainability and the New Economics: Synthesising Ecological Economics and Modern Monetary Theory.* Springer, Chapter 14.

53. Stephanie Kelton's 14-minute TED talk, https://www.ted.com/talks/stephanie_kelton_the_big_myth_of_government_deficits; Alan Kohler's 2.5-minute talk https://www.abc.net.au/news/2020-07-17/what-is-modern-monetary-theory/12455806; and Steven Hail's 1.5-hour lecture presentation https://www.youtube.com/watch?v=qBpm5sVmGYc

54. Steven Hail (2022), *op. cit.*

55. William Mitchell et al. (2019), *op. cit.*

56. Governments borrow money by selling bonds to the private sector investors, such as pension funds and insurance companies.

57. Yeva Nersisyan & L. Randall Wray (2019). *How to Pay for the Green New Deal.* Levy Economics Institute of Bard College, Working paper No. 931, https://www.levyinstitute.org/publications/how-to-pay-for-the-green-new-deal

58. Nersisyan & Wray, *op. cit.*

59. John Maynard Keynes (1940). *How to Pay for the War: A Radical Plan for the Chancellor of the Exchequer.* London: Macmillan.

60. F. Norrestad (2021). Quantitative easing in the U.S.—statistics and facts. https://www.statista.com/topics/6441/quantitative-easing-in-the-us/#topicHeader__wrapper

61. F. Norrestad (2021), *op. cit.*

62. Steven Hail & David Joy (2020). Federal debt and modern money. Global Institute for Sustainable Prosperity. http://www.global-isp.org/policy-note-121

63. Hail & Joy (2020), *op. cit.*

64. Steven Hail (2022), *op. cit.*

65. William Mitchell et al. (2019), *op. cit.*, Chapter 21.

66. J. Beardsley Ruml (1946). Taxes for revenue are obsolete. *American Affairs* 8(1):35–39.

67. Paul Samuelson. The balanced budget superstition. https://www.youtube. com/watch?v=4_pasHodJ-8

68. Steven Hail (2022), *op. cit.*

69. Neoclassical economists attempt to justify these contradictions to their ideology by arguing they are temporary, emergency measures that must not establish a precedent.

70. Paul Samuelson. The balanced budget superstition. https://www.youtube. com/watch?v=4_pasHodJ-8

71. Social justice would require compensation to low-income earners: for example, an increase in the income tax threshold for employees and in pensions or social security payments for unemployed people.

72. IEA. World energy balances highlights. https://www.iea.org/data-and-statistics/data-product/world-energy-balances-highlights

73. Deloitte. *EU carbon border adjustment mechanism.* https://www2. deloitte.com/nl/nl/pages/tax/articles/eu-carbon-border-adjustment-mechanism-cbam.html

74. Social Prosperity Network (2017). *Social Prosperity for the Future: A Proposal for Universal Basic Services.* London: Institute for Global

Prosperity, UCL. https://ubshub.files.wordpress.com/2018/03/social-prosperity-network-ubs.pdf; Anna Coote et al. (2019). *Universal Basic Services: Theory and Practice: A literature review.* Institute for Global Prosperity, UCL. https://ubshub.files.wordpress.com/2019/05/ubs_report_online.pdf; Jason Hickel (2020). *Less is More: How degrowth will save the world.* Penguin Random House, Chapter 5.

75. Boris Frankel (2022). From technological utopianism to Universal Basic Services. In: Samuel Alexander et al. (eds). *Post-Capitalist Futures: Paradigms, policies, and prospects.* Palgrave Macmillan, Chapter 7.

Reference

Weblinks accessed 27/10/2022.

8

Community Action for Social Change

Change comes from power, and power comes from organisation. In order to act, people must get together...Power and organisation are one and the same. (Saul Alinsky)[1]

8.1   Dialogue on Theory and Practice

The scene is a public park in Sydney, Australia. We are sitting on a bench overlooking a small lake. On a tiny tree-covered island, water birds are roosting, taking off, landing and calling out. It seems so peaceful, although we are tense. We are conscious of the horrific war being fought in Ukraine, the possibility that it could escalate to a nuclear war, and the realisation that it is now too late to limit global heating to 1.5 °C. 'We' consists of the authors, Mark and Rod, and a colleague, Li Jing,[2] who has kindly read the draft Chaps. 1, 2, 3, 4, 5, 6 and 7 of this book. We are meeting to discuss the possible content of the final chapters.

*   *   *

© The Author(s), under exclusive license to Springer Nature Singapore Pte Ltd. 2023   **197**
M. Diesendorf, R. Taylor, *The Path to a Sustainable Civilisation*,
https://doi.org/10.1007/978-981-99-0663-5_8

Li Jing: You have presented strategies for transitioning to the Sustainable Civilisation that make sense and are feasible. But to make them work, you need a large number of policies from governments, plus several international agreements. How on Earth will you get governments to implement them?

Rod: That's the problem we struggle with every day. We can solve the problem in theory, but as individuals, we don't have the power to solve it in practice. That is why we've written the book. We are concerned that time is running out and the collapse of civilisation as we know it is likely, yet we still feel there is hope.

Li Jing: So, what's the basis of your hope?

Rod: For a start, not everything depends on governments. Local communities can create their own futures, at least to some degree. They can also push governments to take action on national and international scales.

Mark: Social movements have non-violently toppled dictators and expelled colonial governments. They have gained votes for women and civil rights for people of colour, developed trade unions to protect the rights of workers, stopped the testing of nuclear weapons in the atmosphere, and created and protected magnificent national parks. They have ended government support of slavery and brought soldiers home from war. While we have identified some solutions to problems of ecological sustainability and social justice, other people are thinking, writing and speaking similar ideas. As part of a social movement, we are not alone.

Li Jing: Yet you, may I say 'we', are not the majority. The majority of people assume incorrectly that our crises can be solved with new technologies without any change in lifestyle, society or the economic system.

Mark: Those who control the existing system are very effective in spreading its propaganda, despite the failures of that system. Fortunately we don't have to convince everyone on the planet. Just an energetic fraction of the population to push their governments, business and industry, and the rest of society in the right direction.

Rod: We understand this isn't easy. Governments don't usually act in response to reasoned argument alone. Lobbying a government is almost useless unless it's backed up by a large number of citizens who can potentially remove the government from office. Lobbying a business is almost useless unless it's backed up by the ability to influence its sales or gain the support of its shareholders.

Mark: That's why we need to build a social movement that can challenge the power of corporations and other vested interests and weaken their control of nation-states.

Li Jing: That's hard enough in a nominally democratic system. It's even more difficult when faced with autocratic regimes.

Rod: Yes, autocracies understand the danger to their power and influence from social movements and even quite innocuous community groups, and will move quickly to protect their power. They supress free speech while ignoring environmental destruction. Look at what's happening in Russia where people who challenge Putin are murdered or locked up on trumped up charges.

Li Jing: In my former country, the Chinese communist party understands the dangers of free speech too and has quietly set up organisations to monitor all five official religions in China. It appoints bishops in the Chinese Catholic Church, although the Pope could veto an appointment until recently. Environmental NGOs in China must be registered and be linked to a government organisation; then they must have a party official in a senior role in the NGO.

Mark: This is why part of the campaign to transition smoothly to the Sustainable Civilisation must be to make systems of government more democratic, open and accountable. Unfortunately, the task is not limited to autocracies. Some nominal 'democracies' seem to be moving towards autocracy. The situation in the United States, the cradle of modern democracy, is concerning. The stacking of the Supreme Court with judges of a particular political position is one of the issues; restrictions on voting is another. Nowadays, if the Democrats win an election, the Republicans don't accept the result; if the Republicans win another election, some commentators fear that they will never permit another change of government. [3]

Li Jing: Social change happens and fortunately, it isn't always in the direction favoured by autocrats, capitalists, fascist movements, or polluting industries. Your Chap. 8 must now offer a practical, feasible pathway for communities to push the transition to the Sustainable Civilisation.

Rod: We'll try.

Mark: As a bushwalker, I can see a pathway ahead, although it's goes through some difficult terrain and is indistinct in several places.

8.2   A Strategy for Social Change

Stimulated by the dialogue presented in the previous section, we now set out a broad strategy by which a social movement can drive the transition to the Sustainable Civilisation. Fortunately, we do not have to start from scratch. We can draw upon a substantial body of experience and theory, based on the successful, partly successful and unsuccessful campaigns of the past and present. These campaigns were and are conducted by environmental, social justice, community health, peace, feminist, trade union, civil rights and economic reform organisations, among others. A significant literature is available, from which this chapter distils several key issues. But first, we need a few basic principles, because we do not support all types of social change movement.

What Kind of Social Change?

We seek a peaceful transition to the Sustainable Civilisation without social collapse. War and terrorism are the opposite of this goal. The transition is incompatible with the kind of social change that occurred in the Reign of Terror that followed the French Revolution, the thuggery that helped the Nazis gain power in Germany in the 1930s, the United States government's attempts to stop Salvador Allende being elected president of Chile in 1970, and the imprisonment and forced 're-education' of over a million Uighurs in Xinjiang.

In practical terms, violence by a small minority seeking major social change will simply alienate the majority of the population. This is recognised by powerholders, who sometimes place violent agent provocateurs in peaceful community organisations in order to discredit them.[4] Violence distracts the media away from the issue that must be publicised. Furthermore, violence by community organisations plays into the hands of the people in power who have at their disposal police and military with a much greater capacity to conduct violence.

The kind of social change needed to transition to the Sustainable Civilisation must be based on community participation. It must have deep, extensive roots in local communities. The members of these communities will belong to diverse groups, motivated and united by the goal of a better world, whether it is for the environment, social justice, peace or health. Many of the existing small social change organisations have a non-hierarchical structure in order to foster community participation and group solidarity.

Two Related Strategies

Ultimately, the campaign must be about weakening centralised power structures that have been captured by vested interests, through the power of community action. Those who control the status quo are a tiny minority of people, but most of them are very wealthy. They use their wealth and the connections it brings to capture nation-states as well as some international organisations, such as the World Trade Organisation and the International Monetary Fund, and to dominate the media and social media in order to impose their vision of a civilisation controlled by them. Their purpose is to increase their wealth and power, not to facilitate social improvements.

The strength of a well-organised coherent social change movement is in its number of members, provided they are organised into a coherent force, the diversity of knowledge and skills it can assemble, and their commitment to serve the biosphere and the common good of the people. Ethically this movement is stronger than a self-serving alliance of vested interests and economics ideologues. The key is exercising community power, a dispersed form of power that can be concentrated when needed to strive for unselfish goals through organisation of the people. This is expressed strongly by Saul Alinsky, a professional organiser, in the quotation that begins this chapter.

Social change movements can create models of a better future society on a small scale, without the cooperation of the government: for example, community renewable energy projects; food purchasing cooperatives; business cooperatives; food banks; bulk buying of green technologies; kitchen table discussions among family and friends; child care; parents and teachers associations; trade unions; artists for the environment; and nature protection groups. Occasionally community projects can reach national and international scales. An example is the Grameen Bank, which makes small loans to the poor without requiring collateral. In Bangladesh, where it originated in 1976, it usually lends to landless women, who start a small business and lift their families out of poverty. The Grameen Bank has a repayment rate of 97%. It's an example of 'trickle-up' development that has spread to many other countries. It won the Nobel Peace Prize in 2006.[5]

Community projects can also change national policies and business practices. In the early 1970s, when the Danish government was planning to implement nuclear power, the Organisation for Information about Nuclear Power (Organisationen til Oplysning om Atomkraft, OOA) called for a moratorium while running a grassroots public information campaign across the whole country. At the same time, the complementary Organisation for Renewable Energy (Organisation for Vedvarende Energi, OVE) was set up, recruiting engineers and scientists as well as activists. Building on Denmark's history of small-scale wind power, the positive, constructive, national campaign by OOA and OVE led the public to reject nuclear power and pushed the government to implement policies to support the growth of community wind power.[6] This became the basis for a huge global industry. Other very successful community-based projects can be found among the Right Livelihood Awards.[7]

Despite such successes, transitioning to the Sustainable Civilisation will not be able to rely on grassroots projects alone. Government policies are needed because:

Government approves and often funds major infrastructure such as transmission lines, roads and railways.

Government is responsible for urban planning and land use planning more generally, provided it doesn't simply leave it to the market.

Government is responsible for pollution control, with the same proviso.

Government sets standards and rules, such as energy efficiency standards for buildings, appliances and equipment, and speed limits on roads.

Government is responsible for public facilities, such as public housing, public health, public education, public libraries and public parks.

Government shapes markets (e.g. for electricity and water), sets taxes, fees and tariffs, and determines the operating environment for banks and other financial organisations.
Government funds research and development that industry doesn't support.

All of the above tasks, that are currently carried out by government or delegated by government to business, are relevant to the transition. Therefore, an effective social change movement cannot limit itself to grassroots projects independent of government. It must also work to change government policies at all levels, national, state/provincial and local. Through national governments, it can also create international agreements.

As discussed in the dialogue in Sect. 8.1, lobbying powerholders based on reasoned arguments alone is unlikely to succeed. Therefore, the broad strategy for an effective social change movement must have two principal lines of action: building grassroots projects independent of government and creating a mass movement to change government policies and institutions and to change the practices of business and industry. The two approaches can help each other: growing grassroots projects can contribute to real change and hence the growth of the larger movement, which increases community influence on powerholders to make policy changes to assist more grassroots projects.

8.3   Refuting Myths about Transitioning to a Sustainable Civilisation

The dialogue in Sect. 8.1 raises the question of how we can build the social change movement in the face of propaganda disseminated by vested interests who wish to continue with business as usual, whatever the cost to our planet's life support system and social justice. First, we must refute the propaganda and then communicate widely a broad vision of the pathway(s) to the Sustainable Civilisation.

Chapter 4, Sect. 4.4 refuted the principal myths about renewable energy used by proponents of fossil fuels and nuclear power to slow down the transition to safe, clean, everlasting, renewable energy and energy efficiency. These refutations, together with practical experience with renewable energy, are gradually eroding the energy myths. Chapter 7, Sect. 7.1 refuted several key myths of neoclassical economics. These refutations still need much wider dissemination. Here we sketch replies to some of the myths, objections and obfuscations used to reject proposals to improve the existing socioeconomic system and transition to a sustainable society.

Myth: The Only Possible Alternative to the Present Socioeconomic System Is Communism, Which Has Failed

Even under capitalism, with government by 'the market' (i.e. large corporations or an autocratic who controls the market), a variety of different systems exist. Consider the United States, Finland and China, with very different socioeconomic systems that result in different outcomes in terms of health, education and general wellbeing indicators. The ecological economist Herman Daly envisages a steady-state economy existing under a system of constrained capitalism, where markets are subject to international caps on resource use. Other authors reject capitalism entirely. We authors see it as a question of definition: is an economic system with constrained markets, no growth in the use of energy, materials and land, and no growth in global population, capitalism? Does it matter what we call it?

The Sustainable Civilisation has been developed pragmatically, recognising the damage unrestrained capitalism (i.e. neoliberalism) does to people and the environment. But we do not want to wait for Godot, the complete fall of capitalism. Instead, we see the priority as the need to implement a wide range of desirable government policies and community actions (summarised in Sect. 8.5) while rejecting neoclassical economics and neoliberalism, which are still promoting endless growth and blocking paths to a sustainable future. While this will be difficult to achieve, it is entirely possible. There are clearly many alternatives to the present exploitative system that retain a role for markets. In the United States, Bernie Sanders ran for the Democratic presidential nomination as a Democratic Socialist in 2016 and 2020 and, while he eventually conceded in 2020, he had gained a huge following.

Many proponents of communism would argue that it has not been tried properly yet and that the system that existed under Stalin and his heirs, or exists under Xi Jinping, are travesties of communism. In Cuba, Castro implemented a number of reforms according to communist principles but created an autocracy that permitted no debate on policies.

Myth: Capitalism Is About Reward for Effort

Reward for effort can indeed be a motivating factor for work. However, this observation has been taken to extremes to justify the astronomical disparities in salaries paid to the heads of corporations. In reality, a corporation rests on the collective effort of many people and no individual is worth the huge amounts of money paid to some CEOs.

Many people of the Global South make enormous efforts to survive. Their struggle is not due to lack of effort, but rather to the accident of birth, geographic location, social conditions, the political system that rules them, and exploitation by the Global North and the rich and powerful within their own societies.

Myth: People Are Naturally Competitive and Will Not Accept Greater Government Controls as Part of Transitioning to a More Sustainable Society

This statement contains two assumptions that must be deconstructed. Contrary to the assumption in the first part of that myth, research in anthropology finds that "cooperation and competition coexist in all societies known to mankind".[8] Indeed, cooperation is the basis for every known community. It is also essential for tackling big challenges, whether they are a prehistoric clan hunting a mammoth or the twentieth century project to land people on the Moon. An important factor driving many examples of cooperation is reciprocity: if I contribute something to an individual or a group, I can reasonably expect some benefit in return, either to myself individually or to my group.[9]

The proportions of cooperation and competition are not genetically determined but vary from society to society. Neoliberalism fosters an extreme situation where competition for money, power and status are breaking down communities and impeding cooperative ventures. In particular, *"Part of the problem of diagnosing climate change and considering corrections…is the extreme emphasis that is put on individual independence"*.[10] But even under extreme forms of capitalism, some cooperative enterprises thrive and occasionally have transformed the society and economy of a whole region.[11]

The second assumption implies that the Sustainable Civilisation would necessarily have repressive controls limiting the freedom of the whole population. New laws, taxes and other policies would certainly be needed, but they would be designed to limit the freedom of big business to destroy the planet and exploit the powerless. The myth seems designed to divert attention away from the real threats.

We already have government controls—such as taxes (minimised by the rich), laws to protect corporations, building regulations to protect households, and road rules to protect vehicle passengers and pedestrians. The so-called free market is a myth created and disseminated by businesses and industries to obtain maximum profits while pushing adverse impacts onto the rest of society. The state makes markets work through international treaties, state subsidies, write-offs, legislation to protect corporations, legal waivers, tax holidays, bail-outs, copyright and patent enforcement, investment in public infrastructure, industry friendly fiscal policies, and so on.[12] Where neoliberalism talks about 'freedom of the individual', it's really talking about freedom for a select few. In the Sustainable Civilisation, some of the old controls will be replaced with new controls—such as taxing the rich and replacing part of the large fraction of urban land currently devoted to cars with light rail easements, cycleways and pedestrian areas. Planning, with community input, will play a greater role. Civilisation has grown to such a scale and complexity that an uncoordinated approach cannot work.

Many people would be keen to accept new policies to mitigate the climate emergency, reduce pollution, and facilitate a less exploitative economic system that provides universal basic services and a job guarantee.

Many people would be even keener for a society that's more equal in opportunity, income and wealth, if they knew that they would likely be much healthier on average.[13] Many people would support increased taxes on the rich if they knew about the scientific findings that more equal societies have lower carbon emissions.[14]

The challenge is to communicate the benefits of new policies, and the actions needed to implement them, to the people, against the resistance of powerful vested interests and autocratic rulers.

Objection: A Truly Ecologically Sustainable and Socially Just Society, Such as the Sustainable Civilisation, Is Utopian and Impractical

The same statement can be made about democracy, justice and freedom. In practice, they all fall short of our ideals, but applying their principles has taken us some of the way along pathways towards those ideals and, in some parts of the world, has created better societies than would otherwise exist. At present, a green industrial civilisation, that could be sustained for millions of years, appears impossible, mainly because of limited non-renewable resources, such as phosphorous, that are essential for life (see Chap. 5, Sect. 5.2).[15] However, one that lasts for several millennia may be possible provided urgent effective action is taken to conserve the essentials for life rather than to exploit and trash them. Although falling short of the sustainability ideal, this scenario would be much closer to sustainability than the present civilisation, which could collapse before the end of the twenty-first century.

If a large minority of the people really want it, and another large minority do not object, we could transition to the Sustainable Civilisation. This book sets out the many practical policies and agreements needed within and between countries. Most of them can be achieved under pressure from strong social movements. Indeed, many have already been achieved in a few jurisdictions—the challenge is to build on these precedents to make the policies normal and common throughout the world.

Failure to make this transition is very likely to drive us into a dystopian future.

Myth: Degrowth to a Steady-state Economy Entails Deprivation

The refutation of this myth follows from the previous answers and the discussion of planned degrowth in Chap. 7, Sect. 7.3. The very rich may feel 'deprived' by finding it more difficult and expensive to own and operate a private jet aircraft or ocean cruiser or gigantic mansion. However, most people would be better off in terms of education, health, housing, transport, utilities, product quality, product lifetime and the environment. Planned obsolescence might disappear. They would be more secure in terms of employment and have more satisfying jobs. However, they would have fewer SUVs and overseas trips and less fast fashion.[16] Governments would spend less money on armaments, roads and subsidies to expensive private schools, private healthcare and fossil fuels. They would spend more on public facilities, environmental remediation and the job guarantee. The financial and advertising industries and the military would be smaller. The gap between the rich and the poor would be smaller, resulting in reduced tensions within and between countries.

Much of the reduction in the use of energy and other natural resources would occur at the macro level and would not be felt directly individuals and households, apart from the incentives to purchase greener products. For example, the greater energy efficiencies of EVs, and the low energy inputs per person to cycling, walking and public transport (provided it has high occupancy) would greatly reduce the transportation energy requirement. Oil refineries, which are substantial energy consumers, would become redundant. The widespread use of electric heat pumps, which are much more energy efficient than gas heating, would reduce the demand for heating energy. Countries that previously exported natural gas as LNG would save much energy by closing their gas liquefaction plants.

Myth: Most People Are Happy with Capitalism—Why Would They Want to Change?

First we must convert this myth into a more precise statement: 'most people' should be replaced by 'most middle- and high-income people who are not concerned about environmental destruction or social injustice and are living in a society not being subjected to war, autocratic rule or an epidemic of murder'. Those people are unwilling to change their situation because they are its beneficiaries. But they are a small minority of the world's population and even a minority in some rich countries. High-income earners, in particular, are unlikely to welcome socioeconomic change. However, middle-income earners in many countries are experiencing the increasing prevalence and severity of floods, droughts and wildfires resulting from climate change. They are also experiencing traffic congestion, pollution, crime, job insecurity, high medical costs and inadequate social security.

When asked, most people are not satisfied with the *status quo*. Global public opinion surveys find strong support for increased climate action,[17] even in the USA.[18] The Global Survey for Sustainable Lifestyles surveyed 8000 young adults from 20 countries and found that poverty and environmental degradation are the world's biggest challenges and that young people want to be part of the solution.[19] In a survey on support for the UN Sustainable Development Goals, the top priorities were found to be SDGs 2 (zero hunger), 1 (no poverty) and 3 (good health and wellbeing).[20] Several surveys find strong support for a shorter working week in Global North countries, and trials find evidence of improved productivity, sleep and family life. [21] A job guarantee would benefit individual workers and society.[22]

If the people come to understand the benefits of the Sustainable Civilisation, they will want to go there.

Objection: Degrowth Is Politically Impossible

For decades, the world has been saturated with propaganda that economic growth gives wealth and prosperity and that it can even provide the resources for protecting the environment and eliminating poverty. Although it has clearly failed at the last two tasks and numerous studies have shown that further growth is unnecessary in the rich countries,[23] the growth mantra is still dominant. Therefore, a campaign to promote degrowth directly, or even planned degrowth measured in physical terms as discussed in Chap. 7, would be resisted by almost all governments, businesses and industries.

However, campaigns to reduce wasteful energy demand, to establish a (leaky) circular economy, to terminate state capture by vested interests, to improve health and wellbeing for all, to establish a job guarantee, and to achieve gender equity are likely to be popular with most citizens. In theoretical discussions, the concept of planned degrowth ties all these good things together, but the term is not needed in public campaigns for these specific goals, which can be justified individually. These individual campaigns could succeed in many countries, setting them on paths towards ecologically sustainable, socially just futures. That's the important task, not whether GDP grows or declines.

For people who want to understand the theory, planned degrowth can be justified by exposing the absurdities of neoclassical economics theory (especially the notion of endless growth on a finite planet), the failures of neoliberalism practice (e.g. in the economic response to the Global Financial Crisis and the COVID-19 pandemic) and the injustice of neo-colonialism.

Myth: A Sustainable Alternative to the Present System Is a Society Comprising Small, Independent, Self-sufficient Communities

For brevity, let's call such communities 'earth gardens' after the sustainable living magazine with the same name.[24,25] Earth gardens could play a significant role in the Sustainable Civilisation as a test bed and an educational example. They are also important for rebuilding local communities and thus weakening neoliberalism. Because of the high prices of urban land, most existing earth gardens are in rural areas. The number of future rural earth gardens would be constrained by land availability, given the high global population and its need for food, along with the need to introduce reforestation programs and other sustainable land remediation practices. Unless our planet's human population were substantially reduced, the best way to locate the majority of people is in cities where urban sprawl and environmental impact are contained. Population density and existing urban forms would limit the number and types of urban earth gardens. Earth gardens also depend on an external industrial society for technologies (e.g. solar panels; medications; communication equipment)—they cannot be completely self-sufficient on urban land. Therefore, it seems unlikely that earth gardens would form the basis of a green industrial society although they would be part of it. Another concern is that many existing earth garden communities are inward looking and have very little impact on the population of a country and its governments.

A future 'civilisation' with population, say, of less than one-tenth of the present is the likely outcome of a collapse of the present industrial civilisation—due to nuclear war or climate change or pandemic or another environmental disaster. A society arising under those circumstances is indeed likely to comprise many small rural communities attempting to be self-sufficient, but it could be very different from a utopia of peaceful earth gardens. With the breakdown of governments and the rule of law, some of the new communities would likely become pirates, raiding their neighbours for the limited food, technologies and other resources remaining after the collapse. The result could be many heavily armed, fortified enclaves, pillaging one another. One can only hope that desire and need would eventually lead to peaceful trade and mutual cooperation rather than the violent scenario painted by the *Mad Max* movies.

Obfuscation: Campaigns for Progressive Income Taxes, Wealth Taxes and Inheritance Taxes Are 'Class Warfare'

This is an example of one of the many obfuscations used by vested interests to oppose progressive social change. The spectre of 'class warfare' was the frequent response of former Australian Prime Ministers, John Howard and Scott Morrison, to basic social justice proposals, including opposition to a government policy to reduce the tax rate on high incomes.[26] Sadly, nominally progressive political parties are often intimidated by this kind of obfuscation, mainly because they understand the power of money to control the media and sway public opinion in order to destroy careers.

A contrary view to this myth is that the absence of wealth and inheritance taxes, and the presence of substantial tax concessions for rich individuals and corporations, is the result of undeclared class warfare conducted by the rich.

8.4   Strategy and Tactics

Governments around the world have, to a large degree, focused their policies and structured their institutions to serve powerful corporate and other vested interests, instead of serving public interest (see Chap. 6). Although individuals do not have the power to change these policies and institutions, community groups and social change movements comprising cooperating community groups have already achieved much and can help create a better future environmentally and socially. This section and the following one discuss the strategies and tactics that community NGOs and social movements can take to push for the socioeconomic and political changes needed for the transition to the Sustainable Civilisation. The struggles are fought, non-violently, on two fronts: putting overwhelming community pressure on governments and businesses, and building grassroots projects independent of government.

According to the Chinese general Sun Tzu, author of *The Art of War*, "*Strategy without tactics is the slowest route to victory. Tactics without strategy is the noise before defeat.*" Strategy is the planning and conducting of long-term campaigns to achieve broad goals. It often involves accepting short-term loss for long-term gain. An introduction to strategy in chess[27] can be illuminating even for people who do not play the game. Tactics are the individual steps or tools used in carrying out a strategy.

Tactics can be chosen from writing articles for newspapers, magazines and online websites; preparing other educational materials; gaining coverage in the media and social media; holding public meetings; giving seminars and talks, both face-to-face and online; house to house visits; organising rallies, marches, sit-ins and pickets; shareholder actions; naming and shaming (taking care to avoid defamation); holding strikes and boycotts; taking legal actions (noting that actions against the government and large corporations are rarely successful and can be expensive); creating alternative institutions; and lobbying powerholders (preferably as a representative of a large number of people).

Several books by experienced campaigners on strategies and tactics for nonviolent social change are listed here.[28] Some are classics. The authors include workers, teachers, academics and a young community organiser who subsequently became President of the United States.

Next we summarise the key policies that we, the people, must get implemented by governments, and community actions that we can take without the cooperation of governments.

8.5 Compendium of Key Government Policies and Community Actions Needed

The policies listed below are for national and sub-national governments and are additional to research and education. They are more specific than the broad UN Sustainable Development Goals, which are accepted here with one exception, namely our rejection of further economic growth for the rich countries. To reduce repetition, the community actions are additional to the usual tactics listed in Sect. 8.4, which are very important.

Most of the policies are feasible in countries where democracy is taken seriously. Indeed, some have already been implemented in several democratic countries. The most difficult to implement widely, in our view, have been marked with an asterisk.

The struggle to implement the key government policies needed (and it is a struggle) is already taking place on seven fronts, labelled A to G below. The first two, Parts A and B, have the task of weakening the root causes of the major threats to the natural environment and social justice, namely the power of vested interests and the exploitative economic system they have created. They have not yet received sufficient attention among the public. Part C is the struggle for peace. Parts D–F address climate, energy and material resources respectively. Part G has the goal of ending neo-colonialism. Within each of the parts, the order does not necessarily indicate priority—all the policies and actions are important.

A. *Weakening state capture by large corporations*

Policies required

Publicly funded elections together with ban on large individual or corporate donations to political parties

Legislation to control revolving door appointments

*Removal of legal rights of mining companies to access land against wishes of owners or indigenous custodians

Legal limits on scale of media ownership by an individual or business

Anti-corruption commission

Penalties for politicians who mislead the public

Public register of lobbyists and publication of work diaries of politicians

Legal requirement for politicians and public officials to declare their assets annually

Legal requirement for worker representation on company boards

Rights of workers and trade unions, including right to strike, established in national law and international agreements

Cooperation between those governments of the Global South that are resisting state capture and neo-colonialism

Citizen juries[29] for input into policy development

Direct NGO action

Use media and social media to expose corporate malpractices

Boycott and divest from polluting and exploitative businesses/industries

Invest ethically

Form citizen-based global institutions to sit in parallel with those of the UN

B. *Planned degrowth to a socially just, steady-state economy*

Policies required

*Universal basic services, based on expanding public/social housing, public education, public and community health facilities, public transport, public parks, and public libraries

Planning commission and incentives to establish in-country green industries

Training courses to develop skills in green technologies

*Job guarantee funded by national government, with jobs provided by all spheres of government and registered community NGOs

Shorter working week

Foreign aid from rich to poor countries that assists the development of green technologies while promoting independence

Supplementation of GDP with several indicators of societal and environmental wellbeing

Wealth tax (e.g. inheritance tax and annual property tax) and progressive income tax

Pollution taxes, including carbon tax

Removal of financial incentives for having children

Support for women's reproductive rights, especially family planning

Legislation to strengthen the rights of employees and trade unions, and to require union representation on boards of large corporations

*International agreement to cap extraction of scarce resources

Direct NGO actions

Develop model sustainable communities, both urban and rural

Promote transition towns

Promote simple living

Form and join small business cooperatives

Expose failures of neoclassical economics and neoliberalism

Promote MMT to replace neoclassical macroeconomics

C. *Reducing incidence and severity of wars*

Policies required

Law requiring government decisions to go to war to be debated and approved by Parliament/Congress

Weapons systems for genuine national defence rather than offence

*Amendments to the Nuclear Non-Proliferation Treaty (NPT) to place uranium enrichment and reprocessing of spent nuclear fuel under international control

*Pressure on nuclear weapons states to remove nuclear weapons

Legislated commitment to no first use of nuclear weapons

Signatory to the NPT and the Treaty on the Prohibition of Nuclear Weapons

*National social defence programs, either in parallel with, or instead of, military defence

Government Department of Peace and interdisciplinary peace studies centres in universities

Direct NGO actions

Organise nonviolent action training, including for community social defence

Expand international cooperation between environmental, social justice and peace NGOs, including academics, professionals and religious organisations

Expose arms manufacturers

Promote conversion of military facilities to civilian purposes

D. Energy transition

Policies required

*National and global caps on all fossil fuel mining, with caps declining to zero over time

Tenders or contracts-for-difference with power purchase agreements for large-scale renewable electricity

Funding of major transmission lines and infrastructure of renewable energy zones

Seeding grants for community renewable energy projects

Fair feed-in tariffs for renewable electricity

Energy labelling and minimum energy performance standards for energy-using appliances, equipment and buildings

Mandatory energy audits for sales and rentals of all houses and apartments with results displayed prominently in sales and rental contracts

Reform of electricity markets to foster renewable electricity, energy efficiency and other demand reduction measures

Provision of land for community solar gardens[30]

Carbon price with revenue distributed equally as a dividend to all adult citizens of the jurisdiction

Incentives to households, businesses and industries to replace combustion heating with electrical

Traffic calming, construction of bicycle paths and designation of pedestrian-only zones

Expansion of urban public transport infrastructure and services

Integration of urban and transport planning

Fleet pollution limits on internal combustion (ICE) engine vehicles

Replacement of ICE vehicles in government fleets with electric
Removal of subsidies to production and use of fossil fuels
Membership of international fossil fuel divestment treaty
Assistance to workers in fossil fuel industries to retrain, relocate or transition to retirement
Phase-out nuclear power
Direct NGO actions
Form community renewable energy cooperatives for generation and local distribution of electricity
Form community groups to bulk-buy energy efficient appliances and equipment, renewable energy technologies and electric vehicles
Take nonviolent 'lock-the-gate' actions to impede new fossil fuel mines and nuclear power stations
Promote cycling and walking for commuting

E. *Conservation of renewable resources other than energy*
   Policies required
   Incentives to farmers to practice regenerative agriculture
   Incentives to consumers to choose healthy diets
   Restoration of degraded habitats—including forests, wetlands and waterways—and control invasive species
   Expansion of national parks and marine protection zones and enact stronger laws to protect them
   Legislation to protect biodiversity and especially rare and threatened species
   *Strengthened international Convention on Biological Diversity[31] with strong protocols
   Seed banks and captive breeding of rare and endangered species
   *International agreements to conserve other renewable resources
   Provision of urban land for community gardens
   Direct NGO actions
   Join nonviolent actions to impede deforestation and urban sprawl
   Boycott shops selling timber from unsustainable sources
   Form and join cooperatives for purchasing vegetarian foods
   Set up food banks
   Divest from businesses and industries that fail to conserve renewable resources

F. Conservation of non-renewable resources

Policies required

*National policies and international agreements to conserve non-renewable resources, including caps and taxes on scarce resources with dividends to all citizens

Environmental certification of products made with non-renewable resources

Mandatory biodegradable packaging

Financial penalties for planned obsolescence, unnecessary packaging and products that cannot be reused or recycled

Direct NGO actions

Lobby, picket or boycott shops that supply plastic bags and excess packaging

Reduce consumerism by promoting social valuing of simple living, smaller houses, having fewer consumer goods and spending more time in social activities

Divest from businesses and industries that fail to conserve non-renewable resources

G. Ending neo-colonialism

Policies required

Relevant policies in Parts A–F

*Creditor governments and financial institutions of the Global North cancel sovereign debt of the poorest countries of the Global South or convert it into development finance

In the absence of debt cancellation or conversion, fair negotiations for restructuring of debt owed by Global South to private creditors as well as to governments of the Global North

Global North countries, especially those with monetary sovereignty, fund, without controlling, sustainable development of the Global South

Direct NGO actions

Expand people-to-people international aid for local self-reliance

H. Additional policies and actions proposed by the reader

8.6 Discussion

The task ahead, to transition to an ecologically sustainable and socially just civilisation, is huge. The strategies and tactics can succeed, *given enough time*. However, the campaigns to gain votes for women and civil rights for black Americans, to expel the British colonial power from India, and to abolish the support from slavery by governments, each took at least several decades.[32] Unfortunately, two existential issues that we now face, climate change and nuclear war, could tip civilisation into an irreversible slide into collapse within a single decade. They need the most urgent action.

As discussed in Chap. 4, the technological part of climate mitigation is the least difficult to implement rapidly. In terms of technological progress in renewable energy, we are not starting from time zero—nowadays wind and solar are the least-cost technologies for electricity generation. But, at the time of writing (October 2022), few experts on climate mitigation, let alone powerholders, have recognised that the rich countries must reduce their energy consumption and that this almost certainly requires them to undergo a planned transition to a steady-state economy (SSE) with reduced throughput (see Chap. 7). This would be primarily a hedge against the likely failure of the gamble on the rapid development of large-scale CO_2 removal from the atmosphere.[33] However, it would bring many other environmental and social justice benefits.

The planned transition in the rich countries to an SSE would impact on the economies of the rapidly growing economies which provide many of the manufactured goods for the rich countries. International agreements are needed to ensure that the rich countries assist rapidly growing and low-income countries to build green economies that depend less on exports and, where they do export, focus on 'green' goods. Needless to say, at a time when "*It is well known and deeply concerning that current provision of climate finance for adaptation remains insufficient to respond to worsening climate change impacts in developing country Parties*",[34] it seems very unlikely that rich countries will contribute the much greater resources needed for the sustainable development of low-income countries.

All the policies and actions listed in Part C of Sect. 8.5 would be valuable in reducing the probability of war, but still may not effective in preventing nuclear war between the USA and Russia or the USA and China. Recent events demonstrate that the so-called Balance of Terror does not prevent a nuclear weapons power such as Russia or the USA from invading another country and threatening (explicitly or implicitly) nuclear annihilation if there is a counterattack against the territory of the invading country.

Both these existential threats, climate change and nuclear war, raise the ethical issue: should governments of countries have the right to foster industries and wars that threaten the whole world? Many of us who do not wield political or business power would answer with a firm 'no!'. At least one international precedent supports this ethical position. The United Nations Economic Commission for Europe established the Convention on Long-range Transboundary Air Pollution, which entered into force in 1983 and aims to protect humans and the environment from air pollution that flows across national boundaries.[35] Unfortunately, ethical concerns about climate change and war have received little attention from governments and the mass media.

Notes

1. Saul Alinsky (1971). *Rules for Radicals*. New York: Random House, p. 113.
2. This dialogue is based on conversations with several different people who are subsumed under the name Li Ling. Li means reason, logic, dawn; Jing means quiet, still, gentle, spirit.
3. John Quiggin (2022). https://johnquiggin.com/2022/05/14/the-thirty-nine-steps
4. Steve Chase (2021). *How Agent Provocateurs Harm our Movements*. Washington, DC: ICNC Press; Bill Heid (2011). How to identify an agent provocateur. https://opesr.wordpress.com/2011/08/11/how-to-identify-an-agent-provocateur
5. David Bornstein (1996). *The Price of a Dream*. Dhaka: The University Press Limited and New York: Simon & Schuster; see also https://grameenbank.org/

6. Franziska Mey & Mark Diesendorf (2018). Who owns an energy transition? Strategic action fields and community wind energy in Denmark. *Energy Research & Social Science* 35:108–117. https://doi.org/10.1016/j.erss.2017.10.044

7. Right Livelihood. https://rightlivelihood.org/

8. J.L. Molina et al. (2017). Cooperation and competition in social anthropology. *Anthropology Today* 33(1):11–14.

9. Molina et al. *op. cit.*

10. Mary Catherine Bateson (2016). The myths of independence and competition. *Systems Research & Behavioural Science* 33:674–677.

11. Mondragon, https://www.mondragon-corporation.com/en/; Andrew Bibby (2012). Cooperatives in Spain—Mondragon leads the way. *The Guardian*, 12 March. https://www.theguardian.com/social-enterprise-network/2012/mar/12/cooperatives-spain-mondragon

12. Mark A. Martinez (2009). *The Myth of the Free Market: The role of the state in a capitalist economy.* Kumarian Press; Thomas M. Magstadt (2020). The myth of the free market. *Counterpunch.* https://www.counterpunch.org/2020/01/23/the-myth-of-the-free-market; Robert Reich (2020). To reverse inequality, we need to expose the myth of the 'free market'. https://www.theguardian.com/commentisfree/2020/dec/09/inequality-free-market-myth-billionaires

13. Inequality.org. https://inequality.org/facts/inequality-and-health/

14. Lucas Chancel & Thomas Piketty (2015). Carbon and inequality: from Kyoto to Paris. Paris School of Economics. https://doi.org/10.13140/RG.2.1.3536.0082

15. Although one of us (MD) was originally an astrophysicist, we are not optimistic about schemes to mine the asteroids and moons of the solar system and see no prospect of manned interstellar travel.

16. Ethical Consumer (2021). https://www.ethicalconsumer.org/fashion-clothing/what-fast-fashion-why-it-problem

17. UNDP (2021). https://www.undp.org/press-releases/worlds-largest-survey-public-opinion-climate-change-majority-people-call-wide-ranging-action

18. Pew Research Center (2019). https://www.pewresearch.org/science/2019/11/25/u-s-public-views-on-climate-and-energy

19. UNEP (2011). Global Survey on Sustainable Lifestyles. https://www.unep.org/explore-topics/resource-efficiency/what-we-do/one-planet-network/global-survey-sustainable

20. IISD (2021). https://sdg.iisd.org/news/public-opinion-survey-ranks-global-priorities-for-sdgs

21. Guðmundur D. Haraldsson & Jack Kellam (2021). *Going Public: Iceland's journey to a shorter working week.* Alda & Autonomy. https://autonomy.work/wp-content/uploads/2021/06/ICELAND_4DW.pdf; Bryan Lufkin & Jessica Mudditt (2021). *The case for a shorter workweek.* BBC. https://www.bbc.com/worklife/article/20210819-the-case-for-a-shorter-workweek; Joe Pinsker (2021). Kill the 5-day workweek. *The Atlantic.* https://www.theatlantic.com/family/archive/2021/06/four-day-workweek/619222

22. Pavlina Tcherneva (2020). *The Case for a Job Guarantee.* Wiley.

23. See references on ecological economics cited in Chap. 6.

24. *Earth Garden Journal.* https://www.earthgarden.com.au

25. A similar entity is a 'nowtopia': see Chris Carlsson (2008). *Nowtopia: How pirate programmers, outlaw bicyclists, and vacant-lot gardeners are inventing the future today.* Oakland CA: AK Press.

26. Bernard Keane (n.d.). https://www.msn.com/en-au/news/australia/the-class-war-is-back-and-morrison-and-the-media-are-waging-it/ar-AAXaETp?li=AAgfLCP/; Anon (2019). https://www.sbs.com.au/news/article/shortens-class-war-and-climate-change-cost-labor-john-howard/oyicz7spu

27. Chess.com. https://www.chess.com/terms/chess-strategy

28. Saul Alinsky (1971), *op. cit.*; Gene Sharp (1973) *The Politics of Nonviolent Action*, Porter Sargent; Kim Bobo et al. (2001). *Organizing for Social Change: Midwest Academy for Activists.* Santa Ana CA: Seven Locks Press; Virginia Coover et al. (1978). *Resource Manual for a Living Revolution.* 2nd ed., Philadelphia PA: New Society Press; Mark Diesendorf (2009). *Climate Action: A campaign manual for greenhouse solutions.* Sydney: UNSW Press; Mohandas Gandhi (1927, 1929) *An Autobiography or The Story of my Experiments with Truth*, Ahmedabad: Navajivan Publishing House; George Lakey (1973) *Strategy for a Living Revolution*, WH Freeman; Bill Moyer et al. (2001) *Doing Democracy: The MAP Model for Organising Social Movements.* Gabriola BC, Canada: New Society Publishers; Barack Obama (1995, 2004). *Dreams from my Father.* Melbourne: Text, especially Chap. 12; Chris Rose (2005) *How to Win Campaigns: 100 Steps to Success*, London: Earthscan; Rebecca Solnit (2016). *Hope in the Dark: Untold histories, wild possibilities.* Digital edition. Edinburgh & London: Canongate.

29. NewDemocracy.   https://www.newdemocracy.com.au/what-is-a-citizens-jury
30. Haystacks Solar Garden. https://haystacks.solargarden.org.au
31. Convention on Biological Diversity. https://www.cbd.int/convention
32. The first British campaign against slavery may have been the protest by Quakers in Pennsylvania in 1688. The legislation that finally ended slavery in British colonies was passed in 1843. In the USA, the ratification of the 13th Amendment to the Constitution banned slavery for all except convicts in 1865.
33. Mark Diesendorf (2022). Scenarios for mitigating CO_2 emissions from energy supply in the absence of CO_2 removal. *Climate Policy* 22:882–896. https://doi.org/10.1080/14693062.2022.2061407; Mark Diesendorf (2022). Scenarios for the rapid phase-out of fossil fuels in Australia in the absence of CO2 removal. *Australasian Journal of Environmental Management* https://doi.org/10.1080/14486563.2022.2108514; Mark Diesendorf & Steven Hail (2022). Funding of the energy transition by monetary sovereign countries. *Energies* 15:5908. https://doi.org/10.3390/en15165908
34. United Nations Climate change. *COP26 outcomes: finance for climate adaptation.* https://unfccc.int/process-and-meetings/the-paris-agreement/the-glasgow-climate-pact/cop26-outcomes-finance-for-climate-adaptation#eq-5
35. UNECE. https://unece.org/convention-and-its-achievements

Reference

Weblinks accessed 26/10/2022.

9

Conclusion

Hope locates itself in the premises that we don't know what will happen and that in the spaciousness of uncertainty is room to act. (Rebecca Solnit[1])

Although the situation is dire, the transition, from a failing, collapsing civilisation to one that is ecologically sustainable and more socially just, may still be possible. Our grounds for hope echo Rebecca Solnit's statement (quoted above) that uncertainty about the future offers space for action. We do not know if there will be nuclear war, but we can campaign to try to stop it happening before it is too late. We do not know if widespread grassroots action and proposed government policies can mitigate climate change before we cross a tipping point, but we do know that we have almost all the technologies we need to make this happen. This book and others outline the key socioeconomic and political changes needed to bring us back from the edge. The principal challenges are the huge scope of actions required—from international agreements to national policies to industries and businesses, to towns and to local communities—and the limited time remaining.

Vested interests, whose principal goal is to increase their own wealth and power, have captured many nation-states and slowed down, and in some cases, even reversed the necessary socioeconomic change processes.

© The Author(s), under exclusive license to Springer Nature Singapore Pte Ltd. 2023   **225**
M. Diesendorf, R. Taylor, *The Path to a Sustainable Civilisation*,
https://doi.org/10.1007/978-981-99-0663-5_9

They have distracted us from the key issues with pseudo-scientific and economic myths: myths that humans could live in a world without biodiversity; that variable sources of renewable energy cannot power an industrial society; that endless economic growth on a finite planet is possible and desirable; that wealth trickles down from the rich to the poor; and that superpowers need weapons of mass destruction with the capacity to destroy human civilisation and most of the biosphere many times over. To make matters worse, some environmental organisations, who wish to get a reassuring message accepted by the majority, have spread the myth that technological changes alone will be sufficient to transition to ecological sustainability; that 'green growth' will be safe and effective.

The result of the widespread dissemination of these and other myths is that we are running out of time to make the transition to the Sustainable Civilisation. It is already too late to keep global heating below 1.5 °C and will take a huge struggle to keep it below 2 °C. Planetary boundaries have been exceeded in climate change and losses of biodiversity, freshwater, bio-geo-chemical flows and land use; ocean acidification has almost exceeded the safe boundary. Being human, we must not and cannot surrender, but now that there may be so little time left before the planet crosses a tipping point, we cannot restrict ourselves to relying on the market.

Nevertheless, Table 9.1, similar to a SWOT analysis,[2] offers hope that we could still avoid collapse and transition to the Sustainable Civilisation.

Needless to say, we must implement the strategies discussed in Chap. 4 to transition from fossil fuels to an energy efficient, renewable energy system. And we must implement the strategies discussed in Chap. 5 to reduce the rate of use of renewable resources to their natural rates of replacement and, as far as possible, to substitute renewable resources for non-renewable. But we must do much more and do it urgently and rapidly.

It is essential to weaken the political power of vested interests, especially large multinational corporations with investments in polluting technologies and exploitative businesses and industries. We must replace neoclassical economics with ecological economics and apply Modern

Table 9.1 Assessing the prospect of transitioning to the Sustainable Civilisation

Strengths and opportunities	Threats and weaknesses
Human adaptability and ingenuity	Climate change; pollution; resource depletion; disrupted natural systems
Human capacity for altruism and self-sacrifice	War; nuclear weapons
Energised nonviolent environmental protection and social change movements	State capture; neo-colonialism
Rapidly developing technologies	Growth in human population
The concepts and experiences of democracy, justice, freedom and sustainability	Finite nature of Earth's biosphere and resources
Recent failures of neoclassical economics and neoliberalism	Human propensity for short-term thinking and self-delusion
Growth in understanding of alternative economics systems, including ecological economics, and the concepts of steady-state economics and modern monetary theory	Desire of a subset of humans to dominate through wealth and power: Racism, neo-colonialism, sexism and so on.
Growing awareness of state capture and neo-colonialism	Dominance of capitalism and its ideological prop, neoclassical economics, and residual attachment to its extreme version of practice, neoliberalism

Monetary Theory to help finance the transition. We must campaign on the failures of neoliberalism, when faced by the Global Financial Crisis and the COVID-19 pandemic, to get it rejected entirely by decision-makers and the public, thus weakening the extreme capitalist edifice. We must strengthen the Sustainable Development Goals and ensure they are implemented. New, strong, international agreements on climate mitigation, resource conservation, aid to low-income countries, peace and disarmament must be negotiated. We must urgently gain more signatories to the Fossil Fuel Non-Proliferation Treaty and the United Nations Treaty on the Prohibition of Nuclear Weapons.

All this can be achieved by supplementing specific campaigns—climate action, renewable energy and energy efficiency, resource conservation, environmental protection, social justice, community health and peace—together with over-arching campaigns that address the fundamental driving forces of all these issues. This strategy requires new organisations—together with alliances and networks of existing organisations representing all these issues—to campaign within nations and internationally to publicise and weaken the power of vested interests controlling destructive industries and economic systems and to strengthen democratic decision-making through enhanced community participation. The seeds of this movement are already sprouting.

For example, in the United Kingdom, the organisation Involve[3] develops, supports and campaigns for new ways of involving people in the decisions that affect their lives. Also based in the United Kingdom, with an international outlook, the Foundation for Democracy and Sustainable Development[4] is a think tank that explores and helps build the relationship between flourishing democracy and sustainable development, using research, advocacy and dialogue. The Declaration for American Democracy, a coalition of over 250 member organisations, is campaigning to ensure that all US Americans have the right and ability to vote in elections.[5] The Australian Democracy Network[6] was founded collaboratively by the Human Rights Law Centre, the Australian Conservation Foundation and the Australian Council for Social Services to create "a healthy Australian democracy that puts people and planet first [and brings] people and organisations together to campaign for the changes that make our democracy more fair, open, participatory, and accountable". The European Climate Foundation has launched the Knowledge Network on Climate Assemblies (KNOCA)[7] which brings together policymakers, academics and members of civil society to explore challenges in the design and application of *climate assemblies*. These are new forms of discussion led by citizens, and are increasingly being used in a variety of different countries at different scales of governance to inform policy responses and social action on climate change.[8] The international independent think tank InfluenceMap[9] provides data and analysis on how business and finance are affecting the climate crisis. The International Society for Ecological Economics[10] and its regional societies provide forums for alternatives to neoclassical economics via scholarly journals, books, conferences, workshops and intensive courses. The first university courses on Modern Monetary Theory are being presented.[11] Independent media are providing valuable information on powerful vested interests. A wide range of community NGOs are campaigning for a sustainable society.

The goal is worthwhile—not a utopia, but simply a society that is genuinely democratic, ecologically sustainable, socially just, healthy, peaceful and provides universal basic services and decent work and for all who want it. Although the transition to the Sustainable Civilisation will be difficult, it must be attempted urgently.

The future of humanity depends on it.

For further information and discussion of the ideas in this book, please visit the authors' website for the book, https://sustainablecivilisation.com.

Notes

1. Rebecca Solnit (2016). *Hope in the Dark: Untold histories, wild possibilities*. Digital edition. Edinburgh & London: Canongate.
2. SWOT analysis—that is, assessing the **S**trengths, **W**eaknesses, **O**pportunities and **T**hreats—was developed to assist organisations, usually businesses, to help assess the internal factors (strengths and weaknesses) and external factors (threats and opportunities) that may create opportunities and risks for the organisation. The difficulty of applying it, without modification, to the transition of civilisations is that almost all the threats and opportunities are internal to human societies and behaviour.
3. Involve. https://www.involve.org.uk/our-work.
4. Foundation for Democracy and Sustainable Development. https://www.fdsd.org.
5. Declaration for American Democracy. https://dfadcoalition.org.
6. Australian Democracy Network. https://australiandemocracy.org.au.
7. KNOCA. https://knoca.eu.
8. Foundation for Democracy and Sustainable Development: Climate assemblies. https://www.fdsd.org/ideas/climate-assemblies.
9. InfluenceMap. https://influencemap.org/index.html.

10. International Society for Ecological Economics. https://www.isec-oeco.org.
11. Torrens University Australia. https://www.torrens.edu.au/blog/torrens-university-introduces-world-first-global-online-courses-in-the-economics-of-sustainability.

Reference

Weblinks accessed 26/10/2022.

Glossary

Agrivoltaics Sharing land between agriculture and solar photovoltaics

Anthropocene Proposed new name for current geological epoch reflecting major alterations to Earth systems by human activity

Anthropocentric concept Human-centred concept

Astroturfing The attempt to create the false impression of widespread grassroots support for a policy, individual or product

Autocracy System of government in which a single person or party (the autocrat) possesses supreme and absolute power

Base-load power station Power station designed to operate continuously 24/7 at rated power

Biosphere Region of the surface and atmosphere of the Earth occupied by living organisms

Biodiversity The variety of plant and animal life in the world or in a particular habitat

Biomass Material of biological origin, excluding fossil fuels or peat, that contains a chemical store of energy originally received from the Sun

Bond An IOU for a loan received by a government or company. It provides the bond-holder with a return in the form of fixed periodic interest payments and the eventual return of the principal at maturity

Capitalism An economic and political system in which a country's trade and industry are controlled by private owners for profit, rather than by the state or workers

Circular economy Economy based on the principles of reduce, reuse, recycle, recover, redesign and remanufacture

Climate assembly Discussion led by citizens to inform policy responses and social action on climate change

Combined-cycle gas turbine Power station where the waste heat from a gas turbine is used to produce steam to generate more electricity; it is more efficient but less flexible in operation than an open-cycle gas turbine

Concentrated solar thermal power (CST or CSP) Technology that uses many mirrors to focus sunlight onto a receiver, where a working fluid is heated to high temperature to drive a turbine or heat engine and so to generate electricity

Contestable concept Concept whose meaning emerges from discussion and debate instead a precise, generally agreed definition: e.g. freedom; justice, democracy; sustainability

Contract for Difference (CfD) A long-term contract for low-carbon electricity between a generator and a government. When the wholesale market price for electricity generated by the generator is below the agreed price set out in the contract, payments are made by the government to the generator to make up the difference. However, when the market price is above the agreed price, the generator pays the difference to the government

Corporate capture State capture by corporations

Decoupling Breaking the relationship between economic and physical growth

Degrowth See *planned degrowth*

Demand response Reducing the demand for energy in response to market signals such as time-of-use pricing, incentives or penalties to influence behaviour

Development (of a country or region) Social and economic improvement in a broad sense. In practice, it may involve *neocolonialism*. Also, see *sustainable development*

Ecocentric concept Concept based on a philosophy that places value and importance on the entire environment and all life in it, not just the parts useful to humans

Ecological economics Interdisciplinary field that aims to develop an economic system compatible with ecological sustainability and social justice; different from environmental economics, which is a branch of neoclassical economics

Ecologically sustainable, socially just development Types of economic and social development that protect and restore the natural environmental and social justice

Ecology The science of how organisms interact with one another and with their environment

Economic growth Growth in GDP or GNP

Ecosystem Biological community of living organisms interacting with one another and their physical environment

Electrolysis of water Using direct current (DC) electricity to split water into hydrogen and oxygen

End-use energy or final energy or total final energy consumption (TFEC) Energy supplied to consumers for direct use: e.g. electricity, petrol. It excludes energy losses in the process of converting primary energy to end-use energy: e.g. losses in electricity generation & transmission; oil refineries; compressing or liquefying natural gas

Energy The ability to do 'work' as defined in physics. Energy forms include electrical, thermal, chemical, gravitational, mechanical and kinetic. Measured in joules or watt-hours.

Energy conservation Reducing energy use by accepting a modified or different energy service; mostly involves behaviour change

Energy conversion efficiency The ratio between the useful energy output from an energy conversion device and the energy input into it

Energy efficiency (EE) or efficient energy use Obtaining the same energy service with reduced energy use or more energy services with the same energy use; mostly involves improved technologies; defined as energy output divided by energy input and is usually expressed as a percentage

Energy intensity Energy consumption of a country or region divided by its GDP.

Energy productivity GDP of a country or region divided by its energy consumption

Energy return on energy invested (EROI or EROEI) Energy output from an energy technology or system divided by its life-cycle primary energy input

Energy storage Energy stored in electrical, gravitational, mechanical, chemical or thermal forms

Energy sufficiency See *energy conservation*

End-use energy The part of primary energy—after deduction of losses from conversion, transmission and distribution—that reaches the consumer and is available to provide energy services. End-use energy forms include electricity, heat, and liquid and gaseous fuels.

Final energy See *end-use energy*.

Fossil fuel Natural fuel—coal, oil or gas—formed in the geological past from the remains of living organisms

Gas turbine A type of internal combustion engine that operates with rotary rather than reciprocating motion. It comprises an upstream gas compressor, a combustion chamber and a downstream turbine on the same shaft as the compressor. A jet engine is a gas turbine.

Geothermal energy Energy obtained from heat stored in the earth

Global South Countries and regions where the majority of the population is poor and powerless

Governance Action or manner of governing a state or organisation

Green growth Scenario for economic growth with cleaner technologies

Green hydrogen Hydrogen produced by using renewable energy to split water

Grid Large-scale transmission and distribution system for electricity or a system of power stations connected by transmission and distribution lines

Grid-following inverter Inverter tied to grid frequency

Grid-forming inverter Inverter that enables solar and wind to maintain grid frequency

Gross Domestic Product The total monetary or market value of all the finished goods and services produced within a country's borders in a specific period, normally one year

Heat pump An appliance, usually electrically powered, that circulates a working fluid to carry heat from inside to outside a structure (e.g. refrigerator; air conditioner) or vice versa (e.g. air conditioner on reverse cycle)

Hydro or hydroelectricity or hydro-power Technology that converts gravitational energy of flowing water into electricity

Inflation An increase in the cost of living as the prices of goods and services rise.

Institution An organisation or an established law or practice

Intergenerational equity Equity (equal opportunity) for future generations

Intragenerational equity Social equity, i.e. equity for members of the present generation

Inverter An electronic or mechanical device that converts direct current (DC) electricity into alternating current (AC) that can be fed into an electricity grid or used locally

Job guarantee Jobs funded by government for anyone who wants to work

Less developed country or region See *Global South*

Load Electricity demand on a grid; more precisely, the amount of electricity on the grid at any given time, as determined by demand from households, businesses and industries

Modern Monetary Theory (MMT) A macroeconomic theory that recognises that governments with monetary sovereignty can create debt-free money without financial constraints; however, there are other constraints

Monetary sovereign government A government that issues its own currency, collects taxes in that currency, maintains a floating exchange rate and is not heavily dependent on foreign currencies.

Natural resource Something extracted from nature that can be exploited for economic gain

Neoclassical economics A broad theory that focuses on supply and demand as the driving forces behind the production, pricing and consumption of goods and services. It assumes that individual economic actors have rational preferences, that individuals seek to maximize 'utility' and that decisions are made on the margin. It ignores the roles of culture and institutions in the economy and treats environment as infinite resource and infinite waste bin

Neo-colonialism The practice of granting a sort of independence with the concealed intention of making the liberated country a client- state, and controlling it effectively by means other than political ones

Neoliberal economics or neoliberalism Economic practice that borrows from the theoretical assumptions of neoclassical economics to argue for free trade, low taxes, low regulation and low government spending. In neoliberalism, government exists only to maintain property rights, defend capitalists and maintain price stability

Net energy Energy output minus life-cycle primary energy invested into an energy generating technology or system

Nuclear fission Nuclear reaction in which heavy atomic nuclei are split with the release of energy

Nuclear fusion Nuclear reaction in which light atomic nuclei are fused to form a heavier nucleus with the release of energy

Oligarchy A small group of people having control of a country or organisation

Once-through hydro Hydroelectricity generated from water flowing downhill from either a single dam or no dam

Open-cycle gas turbine See *gas turbine*

Party (to a treaty) A State or other entity with treaty-making capacity that has expressed its consent to be bound by that treaty by an act of ratification, acceptance, approval or accession.

Patriarchy Institutionalised social system in which men dominate over women

Peak-load power station Power station that runs mainly during peak demand periods for electricity. It's much more flexible in varying its output than a base-load power station

Planned degrowth Planned reduction in the use of energy, materials and land, together with stabilisation of the population, with the aim of transitioning to an ecologically sustainable, socially just society; the goal of planned degrowth is an ecologically sustainable, socially just economy achieved in a democratic way

Policy Statement of intention. To be effective, it must be coupled with a strategy and process for implementation

Power (in context of energy) The rate at which energy is converted into work, measured in watts (joules/second) Or, used loosely, another term for electricity

Power-to-X (P2X) Process in which surplus electrical energy is converted to another form or task, such as transport fuels, chemicals, or heat

Primary energy Energy in natural forms—e.g. coal, oil, natural gas, uranium ore, wind, solar—before it's converted to useful final or end-use energy such as electricity or petrol

Prosumer Someone who sells self-generated electricity to the grid and buys electricity from the grid

Pumped hydro Hydroelectricity in which water flows in a circuit between upper and lower reservoirs. Water is pumped uphill using excess electricity and flows down when needed (e.g. during peaks in demand) to generate electricity

Quantitative easing (QE) Governments ask their central banks to create money and use it to purchase government long-term bonds or other financial assets from the private sector. This injects money into the economy

Rebound Increasing energy efficiency results in increased energy use activity; e.g. buying a more fuel-efficient or electric car may result in driving longer distances

Renewable energy (RE) Energy from a source that is not depleted when used, so long as the Sun continues to shine, for example, wind or solar energy. Geothermal energy is usually included although it is gradually depleted

Renewable energy zone Site for multiple large renewable electricity power stations, possibly together with storage, connected to the main grid via a single transmission line

Renewable fuel A gaseous or liquid fuel created from renewable energy, usually renewable electricity

Reverse auction or tender A procurement mechanism by which renewable energy supply or capacity is competitively solicited from sellers, who offer bids at the lowest price that they would be willing to accept. Bids may be evaluated on both price and non-price factors. Winning bids may be offered a *contract-for-difference*

Social defence Nonviolent community resistance to repression and aggression, as an alternative to military forces

Social equity or social justice Equal opportunity. Can be applied between or within countries

Solar garden A community 'garden' for solar electricity, usually used by people who cannot have solar on their own roofs

Solar photovoltaic (PV) power Electricity generated directly by sunlight falling on a special surface on a solar cell

State capture Exercise of power by private actors—through control over resources, threat of violence or other forms of influence—to shape government policies or implementation in service of their narrow interest

Stranded asset An asset that has suffered from unanticipated or premature write-downs, devaluations or conversion to liabilities, e.g. coal-fired power station; camera that uses film

Steady-state economy Economy with constant stocks of people and artefacts, maintained with low levels of energy and materials' throughput that are ecologically sustainable; in brief, an economy with no physical growth; however, within those constraints it is still a dynamic system

Strategy The planning and conduct of long-term campaigns to achieve broad goals.

Sustainability Goal of (ecologically) sustainable development

(The) Sustainable Civilisation The authors' ecologically sustainable and socially just goal for global society

Sustainable development (SD or ESD) Types of economic and social development that protect and restore the natural environmental and social justice (strong definition)

Synchronous condenser Machine that stabilises frequency of alternating current (AC) in the grid, without burning fuel

Synchronous generator A generator of alternating current (AC) driven at the rotational speed that produces the same AC frequency as the grid

System Group of interacting or interrelated elements that act according to a set of rules to form a unified whole; it is more than the sum of its parts.

Tactics Individual steps or tools used in carrying out a strategy

Tender See *reverse auction*

The state The government of a country together with the public service, the military and police that implement and protect its policies

Tipping point Point in time when an Earth system, e.g. climate, changes into a new state that cannot be reversed by human endeavour

Total final energy consumption (TFEC) See *end-use energy*

Universal basic services Provision by government of low-cost basic services, including public health, education, local transport, housing, water supply, parks and libraries

Variable renewable electricity (VRElec) Renewable electricity generation that varies in timescales of minutes to several days; e.g. wind and solar electricity

Index[1]

---

[1] Note: Page numbers followed by 'n' refer to notes.

© The Author(s), under exclusive license to Springer Nature Singapore Pte Ltd. 2023   **239**
M. Diesendorf, R. Taylor, *The Path to a Sustainable Civilisation*,
https://doi.org/10.1007/978-981-99-0663-5